CET Study Guide

CET Study Guide

4th edition

Joseph A. Risse
Sam Wilson

TAB Books

An imprint of McGraw-Hill

New York   San Francisco   Washington, D.C.   Auckland   Bogotá   Caracas   Lisbon   London
Madrid   Mexico City   Milan   Montreal   New Delhi   San Juan   Singapore   Sydney   Tokyo   Toronto

McGraw-Hill

A Division of The **McGraw·Hill** *Companies*

pbk   5 6 7 8 9 FGR/FGR 9 0 0 9 8
hc   1 2 3 4 5 6 7 8 9 FGR/FGR 9 0 0 9 8 7 6 5

Library of Congress Cataloging-in-Publication Data
Risse, Joseph A.
    The CET study guide / by Joseph A. Risse, Sam Wilson.
      p. cm.
    Rev. ed. of: CET study guide / Sam Wilson, 3rd ed. c1993.
    Includes index.
    ISBN 0-07-052933-7 (p).   ISBN 0-07-053022-X (h)
    1. Electronics—Examinations, questions, etc. 2. Electronic technicians—Certification—United States. I. Wilson, J. A. Sam. II. Wilson, J. A. Sam. CET study guide. III. Title.
    TK7863.R57 1995
    621.381'076—dc20   95-35539
                                                        CIP

Acquisitions editor: Roland S. Phelps
Editorial team: Lori Flaherty, Executive Editor
                Andrew Yoder, Book Editor
                Jodi L. Tyler, Indexer
Production team: Katherine G. Brown, Director
                Ollie Harmon, Coding
                Toya B. Warner, Computer Artist
                Jan Fisher, Desktop Operator
Design team: Jaclyn J. Boone, Designer   EL3
             Katherine Stefanski, Associate Designer   0529337

Dedication

As the co-author, I would like to dedicate my work in
writing this book to my grandchildren:
Matthew J. Risse, Stephen V. and Jane E. Armitage,
Kathleen E. and Daniel J. Risse.

Joseph A. Risse

I dedicate my work on this book to my wife, Norma Wilson.

Sam Wilson

Acknowledgments

We also want to thank Roland Phelps and the following
people at TAB/McGraw Hill for help in putting this book together:

Barbara Rubin
Lori Flaherty
Andrew Yoder
Katherine G. Brown
Ollie Harmon
Jan Fisher
Jaclyn J. Boone
and
Katherine Stefanski

Also, to Nandalall Sukhdeo of Bronx, New York
for his timely suggestions.

Contents

Introduction   *xi*

1 **What you should know about the CET test**   *1*
   Important rules for taking the CET test   *8*
   Chapter 1 quiz   *10*
   Answers to Chapter 1 quiz   *11*

2 **Basic math, dc, and ac circuits**   *13*
   Linear, bilateral, two-terminal circuit elements   *13*
   Nonlinear two-terminal components   *26*
   Impedance matching   *37*
   Chapter 2 quiz   *38*
   Answers to Chapter 2 quiz   *46*

3 **Three-terminal and four-terminal components and basic circuits**   *49*
   Basic electric circuits   *49*
   Three-terminal amplifying components   *54*
   Noise in amplifiers   *60*
   Distortion in amplifiers   *62*
   Special ratings   *63*
   Thyristors   *64*
   Practice questions   *64*
   Answers   *72*

4 **Analog circuits**   *75*
   Amplifiers   *75*
   Integrated circuit (IC) operational amplifiers (op amps)   *79*

Basic op amps   *82*
Op amp circuits   *88*
Timers   *93*
Additional circuits you should know   *93*
The phase-locked loop (PLL)   *94*
Power supplies   *97*
Oscillators   *101*
Protection circuits   *101*
Practice questions   *102*
Answers to the practice questions   *111*

5 Digital   *113*

Numbers and counting systems   *113*
Boolean algebra   *116*
Symbols and other identifications of gates   *117*
Important rules for overbars   *123*
DeMorgan's theorems   *124*
Some basics on hardware   *124*
Use of three-state devices   *126*
Flip-flops   *129*
Memories and microprocessors   *137*
An introduction to microprocessor (P) terminology   *140*
Additional terms you should know   *146*
Additional comments on TTL, CMOS, and ECL logic circuitry   *146*

6 Television and VCRs   *149*

Television and FM signals   *149*
Practice questions   *176*
Answers to practice questions   *185*

7 Test equipment and troubleshooting   *187*

Meter movements   *188*
Practice problem   *189*
Oscilloscopes   *193*
Evaluating parameters   *198*
Testing amplifiers   *201*
Practice questions   *204*
Answers to questions   *207*

Practice questions   *208*
Answers to test   *212*

8 Practice for taking CET tests   *213*
The art of test taking   *213*
Practice test   *215*
Answers to practice questions   *230*

9 A time-limit test   *247*
Practice (timed) test #2   *248*
Answers to quiz #9   *268*

Appendices
A Practice for the associate-level CET test   *275*
Instructions for Sections 1, 2, and 3   *275*
Basic mathematics   *276*
Dc circuits   *278*
Ac circuits   *280*
Transistors and semiconductors   *281*
Electronic components and circuits   *286*
Instruments   *289*
Tests and measurements   *292*
Troubleshooting and circuit analysis   *296*
Answers to the practice test   *303*

B Practice test for consumer products   *307*
Section 11: Linear circuits in consumer products   *311*
Television   *314*
Videocassette and video disks   *317*
Troubleshooting consumer equipment   *317*
Test equipment   *320*
Answers to practice test   *321*

C Computer practice test   *325*
Answers to practice test   *337*

D Examples of color codes *339*

E Laws of Boolean algebra *343*

F Acronyms *345*

Index *349*

About the authors *356*

Introduction

BEING A CERTIFIED ELECTRONICS TECHNICIAN (CET) wouldn't mean much if you could get the certificate just for asking or by simply paying a fee. To become a CET, it is necessary to pass a 75-question Associate-level CET test. Passing that test is required of all CETs. A 75-question Journeyman-level CET test in the specialty of the technician must also be passed.

By the time most technicians get ready to take their Journeyman-level CET test, they've been out of school for some time. They need a study guide that allows them to touch up their theory so that they can identify and re-study those subjects that have faded over the years, and brush up on some of the new technology. That is one purpose of this book: to provide a quick, easy reference so that experienced technicians don't have to go into the test cold, without reviewing the subject matter.

This book goes beyond the basic requirements of the Associate-level and Journeyman-level CET tests. For example, we have included more math than you will encounter on your CET tests because we feel that it helps with the understanding of electronics. Do not be put off by this extra material on math. You will not encounter much mathematics on the CET tests that you take.

Another purpose of this book is to provide an overall study review of the material in CET tests. If there are subjects in this book that are unfamiliar to you, it would be a good idea to get one of the many excellent textbooks available from McGraw-Hill and review those subjects in greater depth. If you are a potential Journeyman technician, you have surely been reading the technical magazines. Consult the table of contents in those magazines to find review material to bring your technology up to date.

In some cases, you might find questions in this book with answers that you don't agree with. Please feel free to write to us in care of the publisher and we will gladly discuss those subjects with you. That will help us as much as it will help you. Sooner or later this book will be revised again. We always give those letters serious consideration when we update a book.

When you get ready to take the CET test, send a note to ISCET (International Society of Certified Electronics Technicians) at the following address:

ISCET
2708 West Berry St.
Fort Worth, TX 76109
Phone: (817) 921-9101
FAX: 817 921-3741
FAX: 800 946-0201
E-mail: iscetFW@aol.com

They can supply additional information about the test. Also, they will provide a list of certification administrators in your area.

We never say "good luck" to someone who is going to take a CET test. We think that's an insult because no luck is involved. You are a highly skilled technician and you know that passing the CET test is a matter of knowing your subject. That can only come with study and experience.

Remember, if you don't pass the first time, it doesn't mean that you're not a good technician. It means that you should review one or two technical subjects. After all, you know the importance in keeping your technology up to date.

Always be sure to read the questions in the CET test carefully. Many technicians answer too quickly, without really understanding the question.

When you can say "I'm a CET," we are certain you will feel that it was well worth the effort.

What you should know about the CET test

IN THIS CHAPTER, YOU WILL LEARN ABOUT THE CET TEST for the Associate-level and Journeyman CET ratings. Some helpful hints are given about how to take the test, what material is in the test, and ISCET's rules related to taking the test. Figure 1-1 shows Associate-level and Journeyman-level CET certificates.

The first step in becoming a Journeyman-level CET is to take the Associate-level test. It is not uncommon for a technician to take the Associate-level and Journeyman-level tests together, or at least, within a few days of each other.

This book reviews the range of subjects covered in the Associate-level test, the Consumer Electronics Journeyman test, and the Computer Journeyman test. If you are interested in the Industrial Electronics Journeyman test or any other option, contact ISCET for more information.

■ **1-1A** *Journeyman-level CET Certificate.*

International Society of
Certified Electronics Technicians

Associate Certified Electronics Technician

Expiration Date: ____________ Registration No.: ____________

be it known by these presents that

VOID

has successfully completed the Technical Tests and Requirements giving universal recognition for competence, ability and knowledge as an Associate Certified Electronics Technician.

This certificate is issued jointly by:

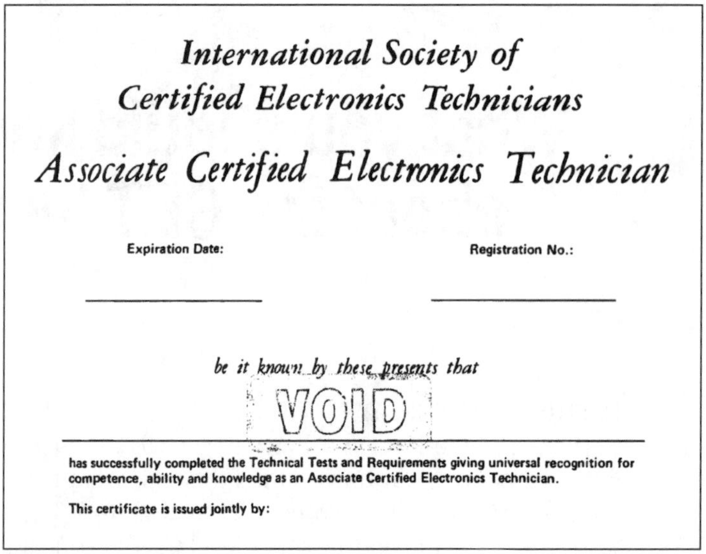

■ **1-1B** *Associate-level CET Certificate.*

Here is a list of the headings of each section of the Associate-level test, and the number of questions that are usually asked under each heading:

Section I		
Basic Mathematics	5 questions	(see Chapter 2)
Section II		
Dc Circuits	5 questions	(see Chapter 2)
Section III		
Ac Circuits	5 questions	(see Chapter 2)
Section IV		
Transistors and Semiconductors	10 questions	(see Chapters 3 and 6)
Section V		
Electronic Components and Circuits	11 questions	(see Chapters 3, 4 and 5)
Section VI		
Instruments	10 questions	(see Chapter 7)
Section VII		
Test and Measurements	10 questions	(see Chapter 7)

Section VIII

Troubleshooting   19 questions   (see Chapter 7)

Total   75 questions

This is a general list. The headings are always subject to change. The range of subjects, however, will be the same, regardless of the headings.

Notice that the Mathematics section has only five questions. This was the most controversial part of the Associate-level test when it was first written. Many technicians claim that they have years of experience without ever using any mathematics. Actually, that isn't possible. They use mathematics every time they make a measurement.

Measurement is an important part of the troubleshooting procedure. For example: A technician makes a measurement of voltage across a resistor. That voltage reading is too low, according to the manufacturer's callout on the schematic diagram. What does a technician know immediately?

Well, he knows that either the current through the resistor is too low or that the resistance has changed to a lower value. How does he know those things? Because he understands Ohm's law, and he knows that the voltage is directly related to the amount of current through a resistor and to the resistance of that resistor. He knows those relationships because he worked many problems in Ohm's law when he was learning basic electronics.

The most important thing about the mathematics for a technician is that it gives an understanding of how the different parameters in the circuit are related. It is highly doubtful that a technician could work efficiently without having done such problems. If it was necessary to learn the relationship between voltage, current, and resistance in a circuit by words rather than by the simple Ohm's law equation, it would have taken much longer to learn how voltage, current, and resistance are related. Suppose, for example, instead of $V = IR$, he had to learn the following expression: "Whenever there is a voltage across a resistor, the magnitude of that voltage is directly proportional to the current through the resistor and also directly proportional to the resistance value of the resistor."

That statement would only explain the relationship of current and resistance to voltage. Another similar statement would be required for current, and another one would be needed for resistance. Can you imagine trying to learn all of the parameter relationships in electronics that way?

Don't be put off by the mathematics section in the Associate-level test. And, don't be put off by the very small amount of mathematics that you will encounter in the Journeyman-level test. The mathematics is only intended to show the relationship between the circuit parameters.

The following material in this book is specifically written to prepare you for the Associate-level test:

Chapters 1, 2, 3, 4, and Appendix A.

All of the material in this book (except Chapter 8 and Appendix C) is related to the Journeyman Consumer test.

Directly related to the Journeyman Computer test are Chapters 5, 6, Appendix C and Appendix E.

After taking the Associate-level test, which is required for all technicians, the next step is to take the Journeyman-level test. A number of different options are available, and the technician will naturally take the test that most closely relates to his four years of experience. Among the options available are:

☐ Consumer electronics (radio, TV, and some audio)
☐ Industrial electronics
☐ Communications
☐ Video electronics
☐ Computer technology
☐ Radar technology

Remember that you need four years of experience in the option that you select. It is possible to get some credit against that four years if you have attended an approved school. If you feel that rule applies to you, write to the ISCET office and ask them for further details. Their address is in the Introduction to this book.

Here is an outline of the Journeyman-level Consumer Electronics CET test along with the usual number of questions in each section:

Section IX

Digital Electronics   8 questions

Section X

Linear (analog) circuits   16 questions

Section XI
Television   15 questions

Section XII
Video cassette recorder (VCR)   11 questions

Section XIII
Troubleshooting   20 questions

Section XIV
Test equipment   5 questions

Total   75 questions

Here is an outline of the Journeyman-level Computer CET test, along with the usual number of questions for each section:

Logic	10 questions
Organization	10 questions
Input equipment	10 questions
Output equipment	10 questions
Memory	15 questions
Programming	5 questions
Troubleshooting	15 questions

Remember that each CET test (Associate level and Journeyman level) is a 75-question, multiple-choice test. That means you will be given a question and number of possible answers. You are to select the correct one in each case.

Be very careful when you are filling in your answer sheet. The answer sheet now being used by ISCET is shown in Fig. 1-2. These tests are graded by machine, so be sure to follow the instructions for giving your answers.

Remember another important point about numbering: When you are taking the Journeyman-level test, the first question is numbered 76. The reason is that numbers 1 through 75 are in the Associate-level test. The separate answer sheet for the Journeyman-level exam always begins with number 76.

No trick questions are used in the CET tests. Sometimes technicians carelessly answer a question and get it wrong. Or, because they haven't thoroughly understood or studied some particular facet of electronics, they get a wrong answer.

-FOR ISCET ASSOCIATE EXAM ONLY-

Information to be Completed to Grade Exams:

Exam Number

Name _____ Date _____ Sex _____ Date of Birth _____
LAST   FIRST   MIDDLE

Name of Examiner _____

Place of Exam _____ State _____ Zip _____

IMPORTANT
USE NO. 2 PENCIL ONLY
● EXAMPLE: ⊂1⊃ ⊂2⊃ ⊂4⊃ ⊂5⊃
● ERASE *COMPLETELY* TO CHANGE
● DO *NOT* FOLD OR STAPLE

PART 1

FEED THIS DIRECTION

FORM NO. 25254-ISCET

SCANTRON

-USE SHADED AREA FOR 25-QUESTION EXAMS ONLY-

Information to be Completed to Grade Exams:

Exam Name _____ Number _____

Name _____ Date _____ Sex _____ Date of Birth _____
LAST   First   Middle

Name of Examiner _____

Place of Exam _____ State _____ Zip _____
For 25 Question Exam, Begin Here ↓   -USE A SEPARATE ANSWER SHEET FOR EACH EXAM-

IMPORTANT
USE NO. 2 PENCIL ONLY
● EXAMPLE: ⊂1⊃ ⊂2⊃ ⊂4⊃ ⊂5⊃
● ERASE *COMPLETELY* TO CHANGE
● DO *NOT* FOLD OR STAPLE

PART 2

FEED THIS DIRECTION

■ 1-2 Two types of answer sheets are used.

When they miss a question, there is a tendency for them to think that it is a trick question; it is not uncommon for a technician to get angry if he believes he has been tricked on a CET test. Let us say this again, and it is very important: No trick questions are used in the CET tests!

Every effort has been made to analyze each question to make sure that it is fair, and ensure that it is a question an experienced technician should be able to answer. Committees check the questions on these tests; they are not the work of one single person. The committees consist of experienced technicians who are now CETs. They have a responsibility to protect the integrity of the CET test; therefore, they don't approve giveaway questions. Also, they do not want questions that are unrelated to the technician's workday experience.

No specialized knowledge of any particular manufacturer's equipment is necessary in order to pass a CET test. You will not, for example, find a question on the startup circuit for an RCA scan-derived power supply. Of course, any technician who works for that manufacturer could quickly answer such a question, but technicians unfamiliar with that particular circuit would not have that specialized knowledge. So, the questions in the CET test are about general knowledge, and about applications that apply to all manufactured equipment.

A two-hour time limit is set for each Associate or Journeyman exam. That is more than ample for a 75-question exam. Always be sure to check your answers for careless errors. If you come to the point when you are no longer making any progress, it is useless to waste your time and the examiner's time by continuing. However, that condition rarely occurs.

Important rules for taking the CET test

These rules apply to both the Associate-level and the Journeyman-level tests. You should read these rules carefully. They are printed on the back of each test. However, you do not want to waste valuable time reading instructions when you are ready to sit for the test.

☐ Do not write in the test book. Use the answer sheet to record your answers. You will need scrap paper to make calculations. The scrap paper will be picked up by the monitor along with your answer sheet.

- [ ] There are no "trick" questions in this exam. Do not look for questions with two or more right answers or questions that have no correct answers.

- [ ] The purpose of the exam is to test your general knowledge of electronics and troubleshooting procedures. Therefore, answer the questions according to what is generally true—not according to some special case that you might know of.

- [ ] If more than one answer appears to be correct, select the one that is most correct without any qualifications. For example, consider this question:

Raincoats are:

1. yellow.

2. waterproof.

3. plastic.

4. obsolete.

To answer this question properly, choice (2) is most correct. It is true that raincoats could be yellow, plastic, or obsolete. However, they are always supposed to be waterproof, so that it is the most correct answer of the choices.

- [ ] You must correctly answer 75% (or more) of the questions to pass any CET exam. Educated guesses are appropriate. In other words, you are not penalized extra points for wrong answers. So, if you cannot decide between choices, pick the best one. Do not leave the answer blank because you will have no chance to get credit for that question.

- [ ] Read and answer each question carefully. Carelessness can cost you valuable grade points.

- [ ] There is a 2-hour time limit.

- [ ] You may use a calculator, provided that it is not programmable, with alphanumeric readouts.

- [ ] No notes or books may be used.

- [ ] No talking is allowed during the test.

When you are ready to take the test, write to ISCET. They will provide a list of certification administrators in your area (be sure to include your phone number when you write). The address of ISCET is given in the Introduction of this book.

Chapter 1 quiz

1. In order to pass a CET test you must answer:

   A. at least 75% of the questions correctly.
   B. at least 70% of the questions correctly.

2. Your first answer for the Journeyman-level test should be put in the space for:

   A. question 76.
   B. question 1.

3. Answer sheets are:

   A. never supplied.
   B. not required.
   C. usually graded by machine.

4. When you are getting ready to take the CET test:

   A. tell the test administrator which type of equipment you work on so the test questions will be related to the equipment you are familiar with.
   B. you will be given a test that does not require a specialized knowledge of any particular manufacturer's equipment.

5. To become a Journeyman CET you must have:

   A. ten years experience in your field.
   B. four years experience in your field.

6. Which of the following statements is correct?

   A. No notes or books may be used when you are taking the CET test.
   B. You may use this book for reference while you are taking the CET test.

7. Which of the following statements is correct?

   A. There is only one correct answer to each question in the CET test.
   B. It is possible to have two or more correct answers for each question in the test.

8. Is the following statement correct? If you can't decide between two statements you should never guess. If your guess is wrong it will cost you extra points on your grade.

   A. The statement is correct.
   B. The statement is not correct.

Answers to Chapter 1 quiz

Question	Answer
1.	A
2.	A
3.	C
4.	B
5.	B
6.	A
7.	A
8.	B

Basic math, dc, and ac circuits

Associate level *This material is related to Sections I, II, and III of the Associate-level test.*

Journeyman level *It would be a good idea to review this material even if you have passed the Associate-level test and you are now preparing for a Journeyman-level test.*

On the subject of electronics, a study of components is usually used first as a foundation for the more complicated subjects. Following a discussion of components, the next subject is usually circuits. *Circuits* are combinations of components. Finally, the circuits are combined into systems which is the most complicated subject in electronics. *Systems* are combinations of circuits.

If you are looking at the overall structure of the CET test, the Associate-level test is primarily concerned with components and the simpler circuits. The Journeyman-level tests have questions about advanced circuits and complete systems. However, this is not a complete delineation because you might find some power-supply systems discussed in the Associate-level test, and you are likely to find questions on components and basic circuits in the Journeyman-level test.

Linear, bilateral, two-terminal circuit elements

Linear circuit components follow Ohm's law. For example, a resistor is a typical linear component. Specific examples are the carbon resistor, the metal film resistor, and the wire-wound resistor. Some other resistors (such as the thermistor and varistor) do not follow Ohm's law, and therefore, are nonlinear.

A circuit component is bilateral if it will conduct equally well in either direction. A resistor is bilateral, whereas a diode is a unilateral component because it conducts in one direction only. Resistors, capacitors, and inductors are the most common linear, bilateral two-terminal circuit elements.

Resistors

The most common uses of resistors in electronic circuits are to:

☐ limit current.

☐ introduce a voltage drop.

☐ generate heat.

All resistors produce a noise voltage known as *thermal agitation noise* (also called *Johnson noise*). That noise exists at all temperatures above absolute zero (–273.16 degrees F). The cause of Johnson noise is random motions of electrons in a resistor. Thermal agitation noise is an example of white noise because it does not have any major frequency emphasis.

This is an important point: Even if a resistor is not connected to anything, it generates a noise voltage across its terminals. If a resistor is connected across the antenna terminals of a receiver, it injects a noise voltage into the receiver input circuit.

So, a resistor is a passive circuit component in most applications, but, since it generates a noise voltage it can also be an active component.

When you purchase a resistor, you must know the resistance value and the power rating. You should also know the voltage rating of the resistor if you are a designer. When some resistors (such as the carbon-composition type) are located in strong electric fields, they undergo a change in resistance. Also, they might be subject to voltage stress between the resistor and the adjacent components.

Carbon-composition resistors

In electronics work, the *carbon-composition resistor* (usually called a *carbon resistor*) is by far the most popular type. The reason for their popularity is easy to understand: It is the cheapest type available today. At one time, you could buy them in ±5%, ±10%, and ±20% tolerances, but the ±20% tolerance resistors are no longer being manufactured. Nevertheless, many of them are still in use.

Before you take the Associate-level CET test, be sure that you know the resistor color codes. They are given in Appendix A. Also, watch for ±1% and ±2% tolerance color codes on newer resistors! See Appendix A.

Carbon resistors can be obtained with power ratings from ⅛ watt to two watts, and they are usually made with a five-color code in the form of bands around the resistors. Experienced technicians who take the Associate-level test have missed questions about color

14

codes for 1- and 10-ohm resistors. To a lesser extent, the 100-ohm resistor color code also seems to cause trouble.

The color code for resistors below 10 ohms is also important if you are planning to take the Associate-level test. You might also encounter questions that require a knowledge of color codes when you take the Journeyman Consumer test.

Practice question

In the circuit of Fig. 2-1, resistor R is part of a power supply decoupling filter. You would expect this resistor to be color coded:

1. brown, black, green, and silver.
2. brown, black, brown, and silver.

Answer: Choice 2 (100 ohms) is correct. A decoupling filter is needed when two or more amplifiers are connected to the same power supply. It prevents the signal of one amplifier from getting into another amplifier through power supply coupling. The resistance value represented by choice 1 (10 megohms) is so large that it would drop most of the supply voltage and starve the transistor amplifier.

In the five-color code, the fifth band is used for reliability information. It shows the number of resistors that must be acceptable in a certain family of resistors. That information is primarily for designers and purchasing agents and has little to do with the technician. However, you should know the purpose of that band.

Carbon resistors have two distinct disadvantages. First, they produce a relatively high noise—especially when compared to some other resistor types available. This noise increases with an in-

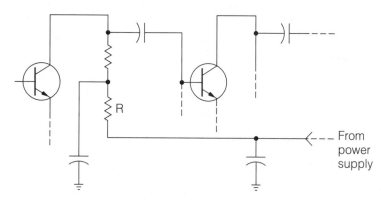

■ **2-1** *Illustration for Practice Question.*

crease in temperature. To be technically accurate, the noise increases as the square root of the temperature increases.

A second disadvantage of the carbon resistors is that they have a relatively high temperature coefficient. This simply means that a small change in temperature can produce a relatively large change in resistance. That action is especially true when the resistor is operated in the 0- to 60-degree Celsius range. A positive temperature coefficient exists for carbon resistors in that temperature range. In other words, the resistance increases as the temperature increases.

Film resistors

Metal-film resistors are more expensive than the carbon type. They are available in resistance values of 0.1 ohm to 1.5 megohm. By comparison, the carbon-composition resistors have values in the range of 1 ohm to over 20 megohms.

Metal-film resistors are normally available in the 0.1- to 1-watt range, and they are less popular than the composition types because they are more expensive. An important feature of the metal-film resistor is its low temperature coefficient. Another important feature is its low noise. Therefore, the two disadvantages of carbon-composition resistors are not a major problem with metal-film resistors.

When comparing the advantages of carbon resistors with metal film resistors it is easy to understand why technicians are cautioned not to replace one kind with another kind. To be sure of making a reliable repair use the same kind of resistor for replacement.

Carbon-film resistors are available in resistance values as high as 100 megohm, and, with tolerances from ± 0.5% up. On the lower end of the available resistance range, carbon-film resistors produce a relatively low noise, compared to carbon-composition resistors. They have a negative temperature coefficient; thus, as the temperature increases their resistance values decrease.

Carbon-film resistors are used most often in places where a very high resistance value is needed or where a negative temperature coefficient is important.

Wire-wound resistors

Wire-wound resistors are more expensive than carbon-composition resistors. They are available off-the-shelf in 0.5- to 3-watt power ratings. Wire-wound resistors are made by winding a resistance wire on a nonconducting form. Normally, winding the wire would cause the resistor to also act as an inductor. However, a special bifilar (noninductive) winding can be used if the self inductance of the component is a serious consideration. An example of a noninductive winding is shown in Fig. 2-2. Wire-wound resistors can be purchased in the 1% to 10% tolerance range. Wire-wound resistors can be purchased with either an inductive or noninductive construction.

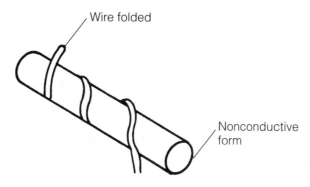

■ 2-2 *Bifilar winding for wire-wound resistors.*

Resistor arrays

Resistors can be purchased in *IC* (integrated circuit) *arrays*. They look like any other IC packages, but inside are series- and parallel-connected resistors. The advantage of the IC configuration is space savings and also the cost savings of using a single package for a number of different purposes.

You are expected to be able to calculate the resistance of series and parallel resistor networks (the same is true for series and parallel capacitors and inductors). Calculation of series and parallel inductors is complicated by problems caused by mutual inductance.

Variable resistors

The most popular variable resistors are either carbon-composition or wire-wound types. These components are connected into the circuit either as *potentiometers* or *rheostats*. The difference between these connections is that a potentiometer is a three-terminal device used to control the voltage of a circuit. A rheostat is a two-terminal component used to control circuit current. The two connections are shown in Figs. 2-3A and 2-3B.

Note: The term *potentiometer* is not only used in reference to a variable resistor used to control voltage. It is also used in reference to an accurate voltage measuring device. Many technicians have

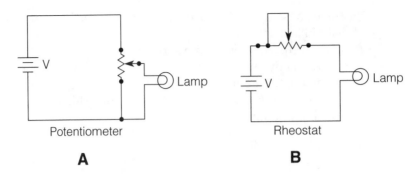

■ **2-3** *Variable resistors are used in two different ways: Potentiometer and rheostat.*

missed questions about potentiometers (the voltage-measuring type) in Journeyman-level CET tests. Those questions are normally found in the Industrial Journeyman CET test.

Capacitors

Capacitors are used in electronic circuits to:

☐ store energy.

☐ introduce a voltage drop.

☐ produce a low-opposition path to high frequencies.

Many technicians have argued that the only use of a capacitor is to store energy. Their argument is very convincing. We list three purposes because they make it easier to understand applications in various circuitry.

At one time, paper capacitors, which used a paper dielectric, were very popular. That type is no longer used extensively for several reasons: The temperature coefficient is very wide and it is difficult to get good insulation between the plates.

Warning: Some paper capacitors have a poisonous material in their dielectric. They should never be taken apart.

The ability of a capacitor to store energy is called its *capacitance*. At one time, it was called *capacity*, but that term is now out of favor.

The capacitance of a capacitor depends on three things:

☐ The area of the plates

☐ The distance between the plates

☐ The dielectric constant of the material between the plates.

Mathematically:

$$C = k \frac{A}{d}$$

where: k is the dielectric constant
A is the area of plates facing each other
d is the distance between the plates

Instead of paper, materials with much higher dielectric constants and much better insulating qualities are now used for inexpensive capacitors. One example is the *Mylar-film capacitor*. Other excellent dielectric materials are polystyrene, polycarbonate, polypropylene, and Teflon. The advantage of the Teflon is its very high voltage-breakdown rating.

Ceramic capacitors are also very popular in modern electronic circuits. Temperature-compensating ceramic capacitors can be purchased with positive or negative temperature coefficients. Also, they are available with a zero-temperature coefficient. The temperature-coefficient rating is identified by a number, such as N220 (which is a capacitor with a negative temperature coefficient) and P750 (which is a capacitor with a positive temperature coefficient). The number (750) shows the number of parts per million change in capacitance with a one degree Celsius change in temperature.

If a capacitor is rated NPO, it has no appreciable change in capacitance for a given change in temperature. When you are replacing a ceramic capacitor, it is very important that you get one with the correct temperature coefficient. If you put a positive temperature coefficient in place of a negative type, it can cause circuit problems. For example, in an oscillator circuit, it can cause the oscillator to drift over a wide range of frequencies for relatively small changes in temperature.

Mica capacitors are used in places where a high dielectric breakdown voltage is important. These capacitors are also useful at high frequencies.

Electrolytic capacitors have a high capacitance in a relatively small package. They are made in two major categories. The aluminum types were the first type sold. They are less expensive than the tantalum type. However, tantalum types have a longer shelf life.

An interesting feature of electrolytic capacitors is their wide range of capacitance values. It is not uncommon for electrolytic capacitor values to range anywhere from 1 microfarad to 100,000 microfarads and more.

Another interesting feature is the wide tolerance ratings of electrolytics. Capacitance tolerances of from –100% to +250% are readily available.

The usual electrolytic capacitors are polarized. If you connect one incorrectly it will be destroyed. Capacitors are now being sold that can store so much energy they are used in place of batteries.

Nonpolarized electrolytic capacitors are available. They can be charged, but they normally have less capacitance than the polarized types.

A nonpolarized capacitor can be made with two identical electrolytic capacitors and two identical diodes. Figure 2-4 shows how it is done.

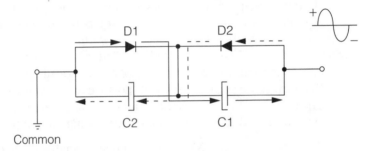

■ **2-4** *Construction of a nonpolarized electolytic capacitor.*

An ac voltage is assumed across the capacitor. On one half cycle the ac voltage is positive. That charges C1 through D1. The electron charge path is shown with solid arrows.

On the next half cycle the ac voltage is negative. During this half cycle, C2 charges through D2. The electron charge path during this period is shown with broken arrows. Nonpolarized electrolytic capacitors are usually purchased as a single unit in one package.

Electrolytic capacitors are usually tested for their ESR (equivalent series resistance). That rating takes into consideration their series and leakage resistances.

Variable capacitors are usually considered to have an air dielectric. In the smaller types, it is common to use some kind of a thin plastic film between the plates to prevent them from shorting out. However, they are still considered to be air-dielectric types. The energy stored by a capacitor is always stored in the dielectric.

In terms of variable capacitors, you should remember that *padders* are used in series with variable capacitors to change the capacitance range. *Trimmers*, on the other hand, are used in parallel with a variable capacitor to change capacitance range. Padders and trimmers are usually small capacitors and they are often mounted on the same frame as larger variable capacitors.

Color codes for capacitors are shown in Appendix D for your convenience. However, it is not typical of a CET test to include questions related to these color codes.

Inductors

Inductors are components that store energy in the form of a magnetic field. By contrast, capacitors store energy in the form of an electric field.

Inductors are used in electronic circuits to:

☐ store energy.

☐ introduce an ac voltage drop.

☐ introduce a high-opposition path to high frequencies.

Here are two important models for the behavior of capacitors and inductors in circuits:

☐ A *capacitor* is a component that opposes any change in voltage across its terminals.

☐ An *inductor* is a component that opposes any change in current through it.

Applications of LCR

Remember that the current in a capacitor leads the voltage across the capacitor by 90 degrees. The current in an inductor lags the voltage across it by 90 degrees.

Compare R, C, and L in the following statements:

☐ The opposition to ac or dc current by a resistor is called *resistance* and it is measured in ohms.

☐ The opposition to ac current by a capacitor is called *capacitive reactance* and it is measured in ohms. It is represented by the symbol X_C.

Mathematically:

$$X_C = \frac{1}{2\pi f C}$$

where: X_C is the capacitive reactance in ohms
π is a constant that is approximately equal to 3.14
f is the frequency in hertz
C is the capacitance in farads

Examples of where you will find capacitors used in electronic circuits are: filter circuits, tuning circuits, and phase-shifting circuits.

The opposition to ac current by an inductor is called *inductive reactance* and it is measured in ohms. It is represented by the symbol X_L.

Mathematically:

$$X_L = 2\pi f L$$

where: X_L is the inductive reactance in ohms
$2\pi f$ is the same as in the X_C equation
L is the inductance in henries

(Note: $2\pi f$ is sometimes called the *angular velocity*. That is a reference to the fact that a sine wave can be thought of as a rotating phasor with an angular velocity equal to $2\pi/T$ where T is the period of the sine wave. But, $T = 1/f$. Substituting $1/f$ for T in the relationship $2\pi/T$ gives $2\pi f$.)

Inductors are used in electronics as part of tuned circuits and filter-circuits. In some circuits, both capacitors and inductors are used to introduce a phase shift or a time delay between voltage and current.

Time constants

Refer to the circuits in Fig. 2-5. Assume that no energy is stored in the capacitor (Fig. 2-5A) or inductor (Fig. 2-5B). When the switch in Fig. 2-5A is turned to position X, the capacitor charges through the resistor. The charging path is shown with a solid arrow.

The time in seconds (T) it takes for the capacitor (C) to charge through the resistor (R) to 63.2% of the applied voltage (V) is called the *time constant* (in some publications, it is represented by τ).

$$T = RC \text{ seconds}$$

The time in seconds (T) it takes for the capacitor to reach full charge—so that the voltage across C equals V is assumed to be five time constants.

Full charge = $5T$ = $5RC$

Now assume the capacitor is fully charged. Refer, again, to Fig. 2-5A. When the switch is turned to position Y, it discharges through the resistor. The path of the discharge electron current is

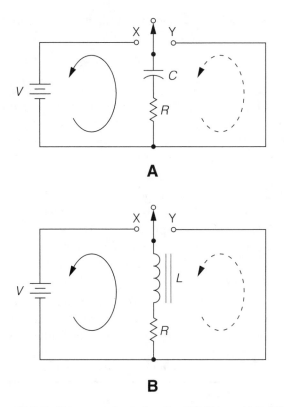

A

B

■ **2-5** *Time constant circuits: RC (A) and LC (B).*

shown with the broken arrow. It takes one time constant (T) for the capacitor to discharge to 36.8% of V.

$$T = RC \text{ seconds}$$

It takes 5 time constants for the capacitor to completely discharge. For the circuit in Fig. 2-5B, the inductor is "charged" when a magnetic field is established around it. When the switch is turned to position X it takes one time constant (T) for the current to reach 63.2% of the maximum current (that maximum current equals V/R). The time constant for RL circuits is calculated as follows:

$$T = \frac{L}{R}$$

It takes five time constants to reach the maximum current.

$$Maximum\ current = 5T = \frac{5L}{R}$$

When the switch is turned to position Y, the current will decrease to 36.8% of maximum in one time constant (T).

$$T = \frac{L}{R}$$

Basic math, dc, and ac circuits

It is considered that no current flows at the end of five time constants:

$$No\ current = \frac{5L}{R}$$

The RL circuit of Fig. 2-5B presents a special problem that is characteristic of all inductive circuits. If the current through the inductor is rapidly changed a very high counter voltage can be generated. Previously it was called the *counter EMF (CEMF)*, but the use of *EMF (electromotive force)* to represent voltage has fallen out of favor.

The counter voltage is generated in the circuit of Fig. 2-5B by operating the switch to produce a very fast increase or decrease in coil (L) current. Despite the fact that we have time constant equations for RL circuits, the high counter-voltages produced makes it difficult to obtain the classic time constant switching delays for current increase or decrease.

The section on the VDR (Voltage-Dependent Resistor) later in this chapter explains how those destructive counter voltages are dealt with. Table 2-1 summarizes the basic equations covered so far.

■ **Table 2-1 Basic equations.**

Ohm's law

$$I = \frac{V}{R} \qquad V = IR \qquad R = \frac{V}{I}$$

Frequency vs. period

$$T = \frac{1}{f} \qquad f = \frac{1}{T} \qquad fr = \frac{1}{2\pi\sqrt{LC}}$$

Impedance (series circuit)

$$Z^2 = R^2 + X_L^2 \qquad Z^2 = R^2 + X_C^2$$

Also,

$$Z = \sqrt{R^2 + (X_L - X_C)^2}$$

Reactance (X)

$$X_L = 2\pi fL \qquad X_C = \frac{1}{2\pi fC}$$

Power

$$P = I^2R \qquad P = VI \qquad P = \frac{V^2}{R}$$

Time constant

$$T = RC \qquad T = \frac{L}{R}$$

See also the color code in appendix D.

Transformers

Transformers are inductive components used in a number of important applications. Figure 2-6 lists the important applications. Special transformers, such as baluns and autotransformers, are used extensively in consumer electronic products. Gyrators are integrated circuits used to simulate inductance in circuits.

Important applications of transformers

* Step voltage up or down.

* Step current up or down.

* Pass ac and reject dc.

* Reduce electrostatic coupling.

* Part of tuned circuit.

* (Along with capacitors) bandpass and bandwidth establish.

* Split phase of ac voltage into two voltages 180° out of phase.

* Convert a grounded ac circuit to a floating ac circuit.

■ **2-6** *Uses of transformers and inductive components.*

Practice question

The dots on the transformer symbol in Fig. 2-7 mean that:

A. the windings must be grounded at those points.
B. the windings must not be grounded at those points.
C. neither choice is correct.

Answer: Choice C is correct. The dots show the points where the voltages across the windings are in phase.

The broken line between the windings in the transformer of Fig. 2-7 represents a Faraday Shield. The transformer windings are made of metal (most are wound with copper wire). Those windings are separated by insulation.

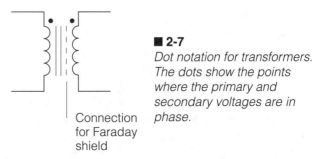

Connection for Faraday shield

■ **2-7**
Dot notation for transformers. The dots show the points where the primary and secondary voltages are in phase.

In some applications, it is important to remember that the combination of metals separated by an insulation makes a capacitor. So, the primary and secondary windings separated by insulation means there is capacitance between those windings.

In some cases, it is important to realize that capacitance can couple energy from the primary to the secondary without using the transformer magnetic coupling. That electrostatic coupling would cause problems in some types of electronic circuits if it wasn't for the Faraday Shield used to prevent it.

Nonlinear two-terminal components

A diode is a popular nonlinear two-terminal component. Both vacuum-tube and semiconductor types of diodes are still in use. In order to stay aligned with the modern technology, there are no questions on vacuum-tubes in CET tests. However, remember that some vacuum-tube diodes must be taken into consideration in the overall electronics picture. For example, a magnetron is a form of a vacuum diode. Magnetrons are used in microwave heating devices.

Breakover devices

The neon lamp is technically not a vacuum-tube device. Instead, it is a gas-filled device. The neon lamp has been replaced in many applications by the diac, a non-linear semiconductor two-terminal breakover device.

The characteristic curves of the two devices (neons and diacs) are very similar (Fig. 2-8). They break over in both directions, so they are bilateral devices.

The neon lamp will not conduct until the firing potential (V_f) is reached or exceeded. The diac will not conduct until the breakover voltage (V_{BO}) point is reached or exceeded. The firing potential of a neon lamp is usually about 65 volts. The breakover voltage of diacs is usually much lower—often less than 10 volts.

Once they break over, diacs and neon lamps have a relatively constant voltage across their terminals, despite changes in current through them. That makes them useful (especially the neon lamp) as simple voltage regulators. Neon lamps are also being used as indicator lights and as simple oscillators. Diacs are used primarily for their breakover characteristics and you will find them in the gate circuits of SCRs and triacs. Both of those devices are discussed in a future chapter. Remember that both the neon lamp and diac are practically nonconducting until their breakover voltage is reached.

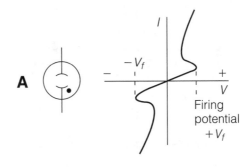

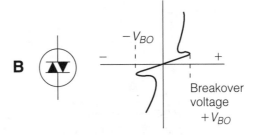

2-8 *The symbol and characteristic curve for a neon lamp (A) and for a diac (B).*

Rectifier diodes

Semiconductor rectifier diodes are used in power supply circuits. In some applications, they might be connected in series or parallel (Fig. 2-9). When they are connected in series (Fig. 2-9A), their peak inverse voltages (PIV) add. The PIV rating gives the maximum allowable reverse voltage across the diode without damaging it.

So, they are series-connected in applications where the PIV rating of one diode is not sufficiently high for the reverse voltage encountered in the circuit.

Resistors are sometimes connected across the series-connected diodes in order to equalize the reverse voltage. That is necessary because the reverse resistances of the diodes are not exactly the same values. Capacitors might also be connected across series diodes to reduce the possibility of damage from transient voltages. Also, the external capacitors equalize the junction capacitances of reverse-biased semiconductor diodes.

Diodes are connected in parallel (Fig. 2-9B) in order to increase their current-delivering capacity. Low-resistance starting resistors are connected in series with the parallel-connected diodes to as-

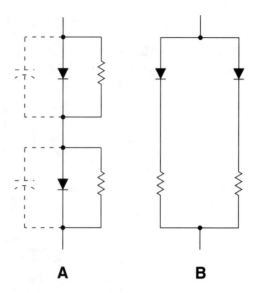

■ 2-9 *Series connection of diodes (A) and parallel connections of diodes (B).*

sure that each diode will conduct and carry its share of the current. Without the low-resistances in series one diode will conduct and shut the other one off. That is known as *current hogging*.

Remember that in semiconductor diodes the side where electrons enter is the cathode. That is always the side marked by the manufacturer. The mark might be a colored band around one end of the diode or it might even be a positive sign. In some diodes, the cathode lead is longer than the anode lead.

Regardless of the method used, the marking is always on the cathode side. Examples of diode color codes are given in the Appendix, but you are not likely to be asked about those color codes.

Constant-current diodes

Constant-current diodes provide a constant-current flow under varying voltage conditions. Figure 2-10A shows the symbol and construction for this diode. One application of constant-current diodes is to linearize the sawtooth waveform in a typical time constant circuit. An example is shown in the UJT (UniJunction Transistor) circuit of Fig. 2-10B.

The UJT is one of a class of semiconducting breakover devices called *thyristors*. The diac is one example. Thyristors are covered in greater detail in the next chapter.

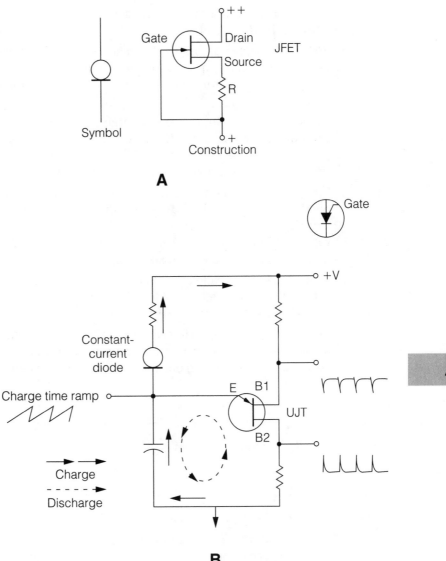

■ 2-10 *The symbol and construction of a constant-current diode (A) with an application (B).*

The UJT is introduced here because it gives an excellent example of how a constant-current diode can be used. During the charge time of the capacitor in the circuit of Fig. 2-10B, the long ramp on the sawtooth waveform is formed.

Normally, that ramp is curved. It has the same shape as a time constant curve. The charging current is linearized by the constant-current diode in the circuit. The solid arrows show the charging path for the RC circuit.

When the voltage across the capacitor reaches a predetermined percent value of the applied voltage (V) the UJT conducts and discharges the capacitor. The discharge path is shown with broken arrows.

The continuous charge and discharge of the capacitor produces the sawtooth waveform at the emitter. A positive-going repeating pulse can also be obtained at base #2 (as shown in the illustration). The negative-going pulse at base #1 is not often used.

The pre-determined decimal value of the applied voltage needed on the emitter for the UJT to conduct is called the *intrinsic standoff ratio*. It is determined by the manufacturer. A popular value is 0.63 because it makes the frequency of oscillation directly dependent upon the RC time constant.

A similar device is called the *PUT (Programmable UJT)*. See Fig. 2-10B. It permits the user to determine the intrinsic standoff ratio by setting the dc voltage on the gate lead.

Figure 2-11 shows the UJT oscillator without the constant-current diode. Observe the curvature on the sawtooth ramp. You have to be careful with the symbol for the constant-current diode. As

30

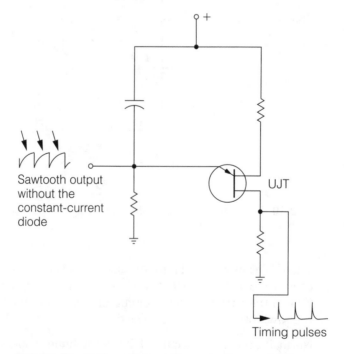

Sawtooth output without the constant-current diode

UJT

Timing pulses

■ **2-11** *The UJT oscillator with sawtooth waveform.*

shown in Fig. 2-12, it looks very much like one of the symbols for a tunnel diode. However, the two devices are nothing alike. Several other symbols for tunnel diodes are shown in the same illustration.

A *tunnel diode* is one of the fastest switches in electronics and switching is one of its most important applications. It is also used as an ultra-high-frequency oscillator.

Optoelectronic devices

Two examples of optoelectronic diodes are the LED (light-emitting diode) and the LAD (light-activated diode). Their symbols are shown in Fig. 2-13. The LED emits light when it is conducting electron current from the cathode to the anode. A cathode-to-anode voltage of approximately 1.6 to 2.0 volts is required for conduction. The manufacturer's recommended voltage is based upon a tradeoff between the amount of light and the life of the diode.

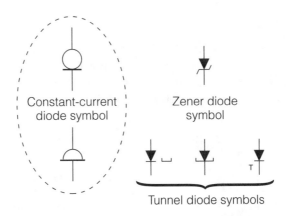

■ **2-12** *Compare the symbol for a constant-current diode with the tunnel diode symbol on the left.*

Light-activated diodes are like open switches in the absence of an input light. When light strikes them, they conduct an electron current in the forward direction (cathode-to-anode), but not in the reverse direction. An optical coupler can use both an LED and an LAD (Fig. 2-13B). Its advantage is a very high isolation resistance between the input and output circuits.

In place of the LAD shown in the illustration many optical couplers use phototransistors or other optoelectronic devices. Remember that the isolation resistance between the input and output circuits

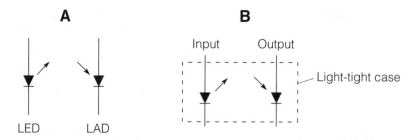

A

LED   LAD

B

Input   Output

Light-tight case

■ **2-13** *Two optoelectronic diodes and their applications in an optical coupler.*

of an optical coupler is extremely high. That makes it possible to interface circuits that operate at two different voltage or power levels.

Optical couplers have also been used as variable resistors in volume-control and tone-control circuitry. They have the advantage that they produce very little noise during changes between the input and output levels.

Other diodes

The four-layer diode (Fig. 2-14) is also known as a *Shockley diode*. Its characteristic curve shows that it is a breakover device in one direction, but it opposes current flow like a rectifier diode in the opposite direction.

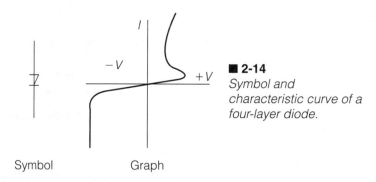

Symbol   Graph

■ **2-14**
Symbol and characteristic curve of a four-layer diode.

Do not confuse Shockley and Schottky diodes. The *Schottky diode* is made by interfacing semiconductor and metal materials at a junction. They are also called *hot-carrier diodes*. Both are represented by the same standard diode symbol.

Hot-carrier diodes are characterized by relatively high conducting-current capability (for their size) and a very low forward voltage drop (see Fig. 2-15). Hot-carrier diodes are used as detectors in receivers.

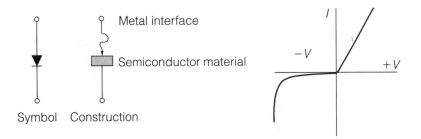

Symbol   Construction

■ **2-15** *Symbol and construction of a hot carrier diode.*

Varactor diodes (Fig. 2-16) behave like capacitors in a circuit. They are operated with a reverse voltage across their terminals. The amount of reverse voltage determines the thickness of the depletion region. That depletion region acts like a capacitor dielectric.

Zener diode symbols   Varactor diode symbols

■ **2-16** *Symbols for the zener diode and varactor diode compared.*

Increasing the reverse voltage also increases the thickness of the dielectric, so, the "plates" of the capacitor are moved further apart. As with any capacitor, moving the plates apart reduces the capacitance. In the varactor diode, the "plates" are actually the P and N regions in the diode. Decreasing the amount of reverse voltage decreases the thickness of the depletion region and increases the capacitance.

Zener diodes, like varactors, operate with the reverse voltage. However, unlike the varactor, the zener diode is operated with a reverse current.

Zener diodes are constant-voltage devices. The voltage across the zener remains constant regardless of the amount of reverse current, provided you adhere to the manufacturer's specifications! The most common application of zener diodes is in voltage-regulator circuits. Figure 2-16 compares the symbols for varactor and zener diodes.

Nonlinear resistors

Thermistors and varistors are important nonlinear resistors. Their symbols are shown in Fig. 2-17. They are not diodes. Their char-

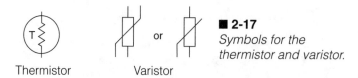

■ 2-17
Symbols for the thermistor and varistor.

Thermistor   Varistor

acteristics are determined by the type of materials used for their manufacture.

Thermistors are temperature-sensitive resistors that are used as heat sensors. Another name for sensor is *transducer*.

Varistors are also called *VDRs*. That is an abbreviation for voltage-dependent resistors. VDRs have a high resistance when there is a low voltage across their terminals, and they have a low resistance when subjected to a high voltage.

Practice question

Why is the VDR connected across the relay coil in the circuit of Fig. 2-18?

A. To reduce the time required for relay contact closure.
B. To protect the transistor.

Answer: Choice B is correct. The input pulse momentarily energizes the relay coil. The inductive kickback resulting from de-energizing the coil would destroy the transistor if the VDR was not in the circuit. The high kickback voltage results in a low VDR shorting resistance to prevent that kickback voltage from destroying the transistor.

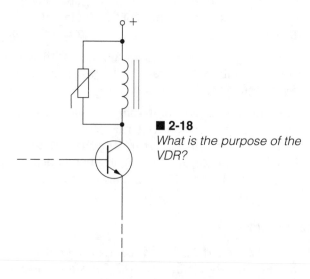

■ 2-18
What is the purpose of the VDR?

Basic math relationships

You should be familiar with the following subjects and equations at the Associate CET Level. Observe that V is used instead of E to represent voltages. That is in accordance with the IEEE recommendations. (IEEE is the Institute of Electrical and Electronics Engineers. One of their functions is to establish standards.)

Be sure that you know how to convert numbers from one radix to another. For example, be able to convert a binary number to a decimal number. That subject is discussed in Chapter 5.

Also, be able to convert between engineering units and powers of 10. For example: Convert 1500 nanofarads to picofarads. (*Answer*: 1,500,000 picofarads.)

Use of complex numbers (numbers with j operators) is a subject you should understand. However, you are not usually required to perform math with complex variables. The complex number for a series RC or RL circuit can be written by inspection. Think of j as being a math operator that tells you to turn right or left.

For an example, see Fig. 2-19A. In that illustration, $6 + j3$ means to go out from the origin (0,0) on a graph a distance of six units and turn left (90 degrees). Then, go a distance of three units.

The complex number $5 - j4$ means to go out from the origin on a graph a distance of five units and turn right (90 degrees). Then, go a distance of four units.

To avoid confusion, the complex impedances are often identified in some way. A dot or a bar over the Z (for impedance) is often used. For the complex numbers in the last examples:

$$\dot{Z} = 6 + j3 \text{ ohms}$$

and

$$\dot{Z} = 5 - j4 \text{ ohms}$$

Remember that inductive reactance is written with a positive j and capacitive reactance is written with a negative j.

Resistances and reactances in complex circuits are in quadrature. That's a fancy way of saying their effects are at right angles to each other. It determines the way you write total currents and voltages.

Practice question

What is the applied voltage in the circuit of Fig. 2-19A?

Circuit Math symbols Graph
          for the circuit

R $R + jX_L$
          $(6 + j3)$

X_L

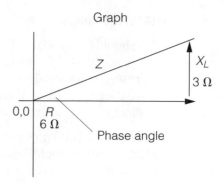

Circuit Math symbols Graph
          for the circuit

R $R - jX_C$
          $(5 - j4)$

X_C

A

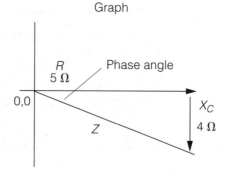

$V = \sqrt{40^2 + 30^2} = 50\ V$

$\phi = \text{Tan}^{-1}\ \dfrac{40}{30} = 53°$

B

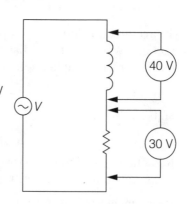

■ **2-19** *Graphical solution for an LC circuit. (Note the phasor marked X_L at right angles to R.)*

Answer: 50 volts. The voltage is calculated by the Pythagorean Theorem.

$$Applied\ voltage^2 = V^2 = [(V_R)^2 + (V_L)^2]$$

First, find the value inside the bracket. Then, take the square root of that value to get the applied voltage.

The phase angle (ϕ) is obtained by using the values of resistance and reactance as shown in Fig. 2-19B.

When complex impedances are written with j operators they are in rectangular form. In polar form the magnitude of the impedance and the phase angle. Here are the rectangular and polar forms of an impedance:

$$\dot{Z} = 3 + j4\ \text{(rectangular form)}$$

and

$$\dot{Z} = 5\angle 53.13°$$

These examples are different forms of the same impedance. The worked problems in complex variables are for your reference only. It is doubtful if you would be required to work complex variable problems like those on a CET test.

Impedance matching

Remember that the maximum power that can be dissipated in a dc circuit occurs when $R_i = R_L$ (see Fig. 2-20). By knowing that theorem, it is easier to understand the need for impedance matching in signal circuits.

All 2-terminal active circuits can be represented by the simple Thevenin circuits in Figs. 2-20 and 2-21. It is demonstrated using a dc network (Fig. 2-20). In an ac or signal circuit (Fig. 2-21), the maximum power dissipation in load Z_L occurs when the internal impedance (Z_i) is the conjugate of Z_L. The conjugate of an impedance is written by changing the sign of the j term as shown in Fig. 2-21.

Additional concepts related to mathematics, dc circuits and ac circuits are covered throughout this book and in the practice tests.

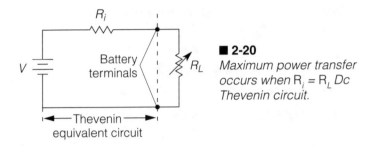

■ 2-20
Maximum power transfer occurs when $R_i = R_L$ Dc Thevenin circuit.

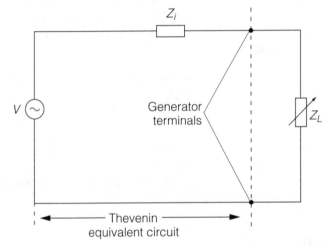

Maximum power transfer occurs when Z_L is the conjugate of Z_i
Examples:
$Z_i = R + jx$, conjugate: $R - jx$
$Z_i = R - jx$, conjugate: $R + jx$

■ 2-21 *Ac Thevenin circuit.*

Chapter 2 quiz

1. A thermistor has a wide variation in resistance corresponding to a relatively narrow change in:

    A. time.
    B. temperature.
    C. humidity.
    D. voltage.

2. In order to be able to calculate the amount of current through a resistor by Ohm's law, it is necessary for that resistor to be:

    A. a VDR.
    B. a thermistor type.
    C. linear.
    D. unilateral.

3. Which of the following is not a linear resistor?

   A. Film
   B. Carbon composition
   C. VDR
   D. Wire wound

4. A certain resistor has a resistance value that depends on the amount of light striking it. It is called:

   A. a varactor.
   B. a PER.
   C. a PIR.
   D. an LDR.

5. In order to show that a resistor has a tolerance of ±10 percent:

   A. the third band is silver.
   B. the fourth band is gold.
   C. the fourth band is silver.
   D. no color is in the fourth band.

6. Select the proper color code for a 100-ohm resistor.

   A. brown, black, brown
   B. brown, black, black
   C. brown, black, gold
   D. brown, black, silver

7. Which of the following is not a common taper for variable resistors?

   A. Linear
   B. Audio
   C. Converse
   D. Reverse

8. A variable resistor with a linear taper:

   A. has a code letter L on the case.
   B. is always wire wound.
   C. can be identified with an ohmmeter measurement.
   D. has a red dot near the center terminal.

9. To stabilize against temperature changes, a circuit can use:

   A. a thermistor.
   B. an LDR.
   C. a ferrite bead.
   D. a varistor.

10. Identify the polarized capacitor in the following list:

   A. metalized paper
   B. glass
   C. ceramic
   D. electrolytic

11. A certain capacitor has a rating of P150. This means that:

   A. the capacitance will decrease 150 parts per million when the temperature increases 1 degree Celsius.
   B. the value of capacitance is 150 microfarads.
   C. the value of capacitance is 150 picofarads.
   D. None of these choices is correct.

12. Which of the following is an advantage of mica capacitors?

   A. Available in both low- and high-capacitance values
   B. High breakdown voltage

13. Variable capacitors can be purposely made nonlinear (by controlling the shape of the plates) in order to get a linear dial on the radio. Is this statement true or false?

   A. True
   B. False

14. Which of the following is not an advantage of tantalum over aluminum-oxide electrolytics?

   A. Longer shelf life
   B. Much lower cost

15. A measure of capacitor leakage resistance is:

   A. ESR.
   B. Vdcw.
   C. Q.
   D. specific dielectric capacitance.

16. The reciprocal of dissipation factor for a capacitor is:

   A. ESR.
   B. Vdcw.
   C. specific dielectric capacitance.
   D. Q.

17. When two capacitors are placed in parallel, the breakdown-voltage rating of the combination is equal to:

   A. the lower of the two breakdown-voltage ratings.
   B. the total of the breakdown-voltage ratings of each capacitor.
   C. a value that is proportional to the reciprocal of the sums of the breakdown-voltage ratings of each capacitor.
   D. the average of the two breakdown-voltage ratings.

18. Three parallel resistors are each dissipating three watts of power. What is the power dissipation of the circuit?

    A. 1 watt

    B. 3 watts

    C. 9 watts

    D. Cannot be determined from the information given.

19. Two 5-microfarad capacitors are connected in series. Their combined capacity is:

    A. 2.5 microfarads.

    B. 3.3 microfarads.

    C. 5 microfarads.

    D. 10 microfarads.

20. A certain resistor is color-coded orange, orange, orange, gold. By actual measurement, its resistance is 34.8 kilohms.

    A. The resistor is out of tolerance.

    B. The resistor is in tolerance.

21. Two coils have equal lengths and equal radii. They are made with slightly different wire sizes. Which of the following is correct?

    A. They will both have the same inductance value.

    B. The one with the larger wire size will have the larger inductance value.

    C. The one with the smaller wire size will have the larger inductance value.

22. The inductance of a coil does not depend on:

    A. the number of turns in the coil.

    B. the current flowing through the coil.

    C. the distance between the turns of the coil.

    D. the shape of the coil.

23. The capacitance of a four-microfarad capacitor in series with a six-microfarad capacitor is:

    A. 2.4 microfarads.

    B. 3.1 microfarads.

    C. 10 microfarads.

    D. 24 microfarads.

24. The capacitance of a capacitor is not affected by the:

    A. type of material used for the dielectric.

    B. area of the plates facing.

    C. type of metal used for the plates.

    D. distance between the plates.

25. The efficiency of a power supply that is delivering maximum power is:

    A. minimum.
    B. maximum.
    C. 50%.
    D. 0%.

26. The maximum power that a battery can delivery to a load resistor occurs when:

    A. a short circuit exists across its terminals; that is, when RL = 0.
    B. an open circuit exists across its terminals. In other words, the output resistance is infinitely high.
    C. the load resistance is adjusted to equal the internal resistance of the battery.
    D. cannot determine the answer from the information given.

27. Electron current flows in a pnp transistor:

    A. from emitter to collector.
    B. from collector to emitter.

28. Doubling the dc voltage across a thermistor will cause the current to:

    A. double.
    B. be halved.
    C. change, but it is not possible to determine the current value by Ohm's law because a thermistor is a nonlinear component.

29. A certain resistor changes its resistance value in accordance with the amount of voltage across its terminals. It is called:

    A. a VDR.
    B. a VER.
    C. a VIR.
    D. an EDR.

30. The resonant circuit of Fig. 2-22 is tuned by "knifing the plates." In other words, the plates of C are moved closer together or farther apart to change f_r. Which of the following will raise the resonant frequency?

    A. Moving the plates closer together
    B. Moving the plates further apart

42

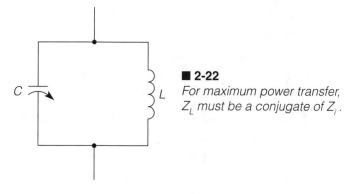

■ **2-22**
For maximum power transfer,
Z_L *must be a conjugate of* Z_i.

31. Figure 2-23 shows an oscilloscope display of a sine-wave signal. What is the frequency of the sine-wave signal?

    A. 62.5 Hz
    B. 156.25 Hz

32. What is the RMS value of the sine-wave signal in Fig. 2-23?

    A. About 11.3 + volts
    B. About 5.66 volts

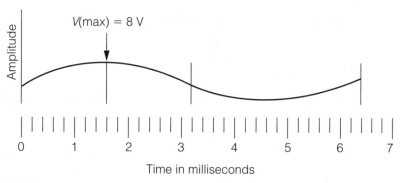

■ **2-23**

33. Convert 100 nanoseconds to microseconds. The value in microseconds is:

    A. 150.
    B. 1500.

34. On an oscilloscope display of a pure sine wave, the peak-to-peak value is 7 volts. The average value of the displayed voltage is about:

    A. 6.36 V.
    B. 3.18 V.

C. 0

D. 2.23 V.

35. Diodes are sometimes connected in series to get:

A. a higher peak inverse voltage (PIV) rating.

B. a higher current rating.

36. Increasing the inductance a series LC circuit will:

A. increase the resonant frequency.

B. decrease the resonant frequency.

37. A semiconductor diode has a white band around one end. That means:

A. it has a current rating of 9 amps.

B. it has a PIV rating of 9 volts.

C. the cathode lead is at that end.

D. the anode lead is at that end.

38. The capacitance of an air-dielectric capacitor will be increased by:

A. moving the plates closer together.

B. moving the plates further apart.

39. A varactor diode is normally operated with:

A. a forward voltage.

B. a reverse voltage.

C. no voltage.

40. Refer to Fig. 2-24. It shows resistors in an integrated circuit array. The resistance between A and B is _____ ohm(s).

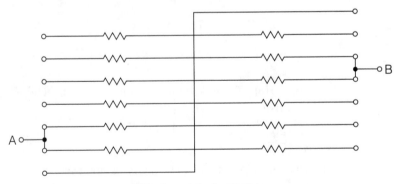

Each resistor is 30 Ω

■ 2-24

41. Which of the following diodes is normally operated with a reverse current?

  A. Zener diode
  B. Varactor diode

42. What is the maximum output power from a 12-volt battery that has an internal resistance of 6 ohms?

  _____ watts

43. Convert 150 nanoseconds to microseconds. The answer is:

  A. 1.5 microseconds.
  B. 1500 microseconds.
  C. 0.015 microseconds.
  D. None of the choices is correct.

44. When a power supply is delivering its maximum power to a resistor, the efficiency of the circuit is:

  A. 100%.
  B. 50%.
  C. 0%.

45. Which of the following is the same as 1000 microvolts?

  A. 0.1 millivolts
  B. 1 millivolt
  C. 10 millivolts
  D. 100 millivolts

46. The reactance of a 10-mH choke at a frequency of 500 Hz is approximately:

  A. 15,750 ohms.
  B. 31,500 ohms.
  C. 50,000 ohms.
  D. None of these choices is correct.

47. A circuit contains a 50,000-ohm resistor and a 25,000-ohm resistor. They are connected in series and 18 volts is applied across them. How much voltage is dropped across the 50,000-ohm resistor?

  A. 3 volts
  B. 6 volts
  C. 12 volts
  D. 15 volts

48. If a sine-wave voltage measures 220 volts RMS, its peak-to-peak value is approximately:

    A. 220 volts.

    B. 440 volts.

    C. 315 volts.

    D. None of these choices is correct.

49. In a simple dc circuit, if the resistance stays the same, and the current increases, it usually means that the voltage:

    A. has increased.

    B. has decreased.

    C. has not changed.

Answers to Chapter 2 quiz

Question	Answer
1.	B
2.	C
3.	C
4.	D
5.	C
6.	A
7.	C
8.	C
9.	A
10.	D
11.	D
12.	B
13.	A
14.	B
15.	A
16.	D
17.	A
18.	C
19.	A
20.	A
21.	A
22.	B
23.	A
24.	C
25.	C
26.	C
27.	B
28.	C

Question	Answer
29.	A
30.	B
31.	B
32.	B
33.	A
34.	C
35.	A
36.	B
37.	C
38.	A
39.	B
40.	30 Ω
41.	A
42.	6
43.	D
44.	B
45.	B
46.	B
47.	C
48.	D
49.	A

47

Three-terminal and four-terminal components and basic circuits

3

Associate level *Important review material for Sections IV and V of the Associate-level CET test.*

Journeyman level *Important review material for the Consumer Journeyman test for Section XI and for other Journeyman tests.*

Chapter 2 is a review covering some of the characteristics of non-electronic passive components (such as resistors, capacitors, and inductors) and some of the electronic two-terminal devices.

49

This chapter covers some three-terminal amplifying and switching components and also some basic circuits for two-terminal linear components. Be sure to understand the operation and circuitry for three-terminal thyristors (such as the UJT, SCR, triac, and three- and four-layer diodes).

Basic electric circuits

You should review some combinations of passive two-terminal linear bilateral circuit components before taking the Journeyman-level CET test. One example is passive circuits, like filters. Remember that passive circuits do not generate a voltage. Voltages are generated in active circuits.

Experts agree that you cannot generate a current. You might run across the expression "generated current." That term is defined as a current that flows as a result of a generated voltage.

Figure 3-1 shows four basic four-terminal filter configurations that use passive components. Remember that inductors pass low frequencies and reject high frequencies. Capacitors pass high fre-

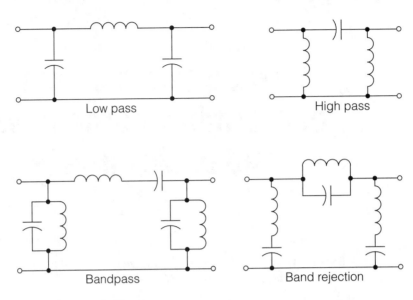

■ 3-1 *Four basic filter types.*

quencies and reject low frequencies. It is an easy matter to identify which type of filter you are looking at. For example, in the low-pass filter the inductor is connected in series between the generating end (left side) and the load end (right side). It will pass the low frequencies from the generator to the load. At the same time, the capacitors that are across the line bypass the high frequencies to ground.

All of the filters in Fig. 3-1 are examples of four-terminal networks. With the common points tied together, they are also examples of three-terminal networks. Those definitions only affect the way the circuits are treated in a mathematical analysis.

The characteristics of series- and parallel-tuned circuits are also important. They are summarized in Fig. 3-2. That illustration shows that in a parallel-tuned circuit, the resonant frequency can be selected by adjusting the variable resistor in either of the branches! In this sense, parallel-tuned circuits are different from the series-tuned circuits that cannot be frequency-adjusted by using a resistor.

The equations for series and parallel resonant frequency and the characteristic curves for those tuned circuits are shown in the illustration. Notice that in the parallel resonant equation, if RL and

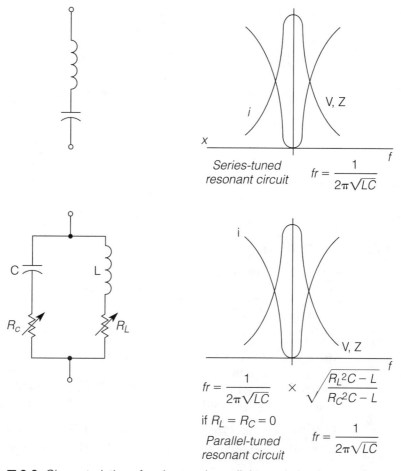

$$fr = \frac{1}{2\pi\sqrt{LC}}$$

Series-tuned resonant circuit

$$fr = \frac{1}{2\pi\sqrt{LC}} \times \sqrt{\frac{R_L^2C - L}{R_C^2C - L}}$$

if $R_L = R_C = 0$

Parallel-tuned resonant circuit $\quad fr = \frac{1}{2\pi\sqrt{LC}}$

■ **3-2** *Characteristics of series- and parallel-tuned circuits.*

RC can be considered to be zero ohms, the equation is the same as for a series-tuned circuit.

The parallel circuit with an RC and/or RL connection is not very often taken up in basic theory books. So, the concept of resistance tuning is unfamiliar to many technicians. However, traps have been designed that are resistance tuned. (It is not necessary to memorize the general-case equation for parallel-tuned circuits.)

Time-constant circuits, like those shown in Fig. 3-3, are also very important if you are going to take a CET test. They might not be shown as individual circuits as they are in this illustration. Instead, they might be part of a more complex circuit.

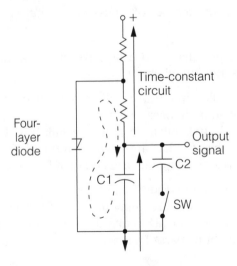

+V

R

$T = RC$

C

+V

R

$T = \dfrac{L}{R}$

L

■ **3-3**
*RC and RL time
constant circuits.*

Practice question

Refer to the circuit in Fig. 3-4. When capacitor C2 is switched into the circuit the frequency of the oscillator will:

A. increase.
B. decrease.

Answer: Choice B is correct. When capacitors are combined in parallel, the capacitance is increased. Increasing the capacitance will also increase the time (T) for one cycle. Because $f = 1/T$, it follows that increasing the time for one cycle will reduce the frequency.

Resistive networks that are used to match impedances are called *attenuators* or *pads*. In some publications, the pads are shown

+

Time-constant circuit

Four-layer diode

Output signal

C1

C2

SW

■ **3-4** *What is the effect of C_2?*

with fixed resistors and the attenuators are made with variable resistors. That is not a hard and fast rule.

Attenuators and pads can be used to match a higher impedance to a lower impedance. In that sense, they perform the same job as a transformer. However, a greater loss will occur in attenuators and pads.

That loss might be desirable and it is called the *insertion loss*. For instance, you might wish to match a high-amplitude signal to a circuit that cannot tolerate the high-amplitude input. Examples are antennas that are mounted in strong-signal areas that tend to overdrive the receiver on one channel. The pad can be switched into the circuit to reduce the amplitude of that signal by introducing the required insertion loss into the line. If the pad is properly designed, it will not cause any impedance mismatches.

One circuit that is sometimes asked about in the consumer CET test is illustrated in Fig. 3-5. This circuit is called a *safety circuit*. It is used in transformerless types of receivers and test equipment. Its purpose is to prevent the possibility of electrocution in the event that the line plug is inserted into the ac power socket in such a way that the chassis becomes hot (above ground).

The need for a safety circuit is eliminated if the ac plug is polarized. However, there is no guarantee that the customer will not at some time replace the plug, despite warnings from the manufacturer. So, the safety circuit is still a good idea. Transformer-type power supplies do not require safety circuits.

A problem with this safety circuit is that (unfortunately) it is never (or almost never) checked by a technician when servicing the equipment. Actually, it should be tested and checked every time the equipment is in the shop for repair. When this safety circuit is defective it does not prevent the operation of the receiver. Always be sure to test this circuit!

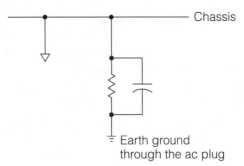

■ **3-5** *A safety circuit for ac/dc equipment.*

53

Three-terminal amplifying components

Do not attempt to take the Associate-level or Journeyman-level CET test unless you are completely familiar with all of the dc operating voltages that are required on all electrodes of the amplifying components (see Fig. 3-6). It would certainly be a disadvantage to try to troubleshoot electronic equipment if you do not know the typical operating voltages and the polarities of voltages required for operating these components.

In the ISCET tests for Associate-level, Journeyman Consumer, and Journeyman Computer, there are no questions on vacuum tubes. Questions about solid-state, three-terminal components *are* used on the ISCET tests.

The polarities of voltages around each component in Fig. 3-6 are meant to show only the voltages with respect to the emitter (E) or source (S). The voltages on those electrodes might be different with respect to other points in a circuit.

You might be asked to identify the polarities of voltages between the electrodes. That type of question is demonstrated in the following question.

Practice question

In an operating NPN transistor amplifier the base is:

A. positive with respect to the collector.
B. negative with respect to the collector.

Answer: Choice B is correct because the collector is positive with respect to the base. That means the base is negative with respect to the collector. There might be an exception to this rule when the transistor is being used as a switch.

Another reason for studying the illustrations in Fig. 3-6 is to be certain that you understand and have memorized all of the symbols of all of these amplifying components. In the past technicians have had trouble identifying the depletion versus the enhancement-type MOSFETs. They have also missed questions about the difference between the N-channel and P-channel device types. Remember this important fact: the arrows always point to the N region in electronics symbols! Therefore, if the arrow is pointing toward the channel, it is an N-channel device.

It is presumed that you know the classes of amplification and these will not be reviewed here. Make sure you know them before you go

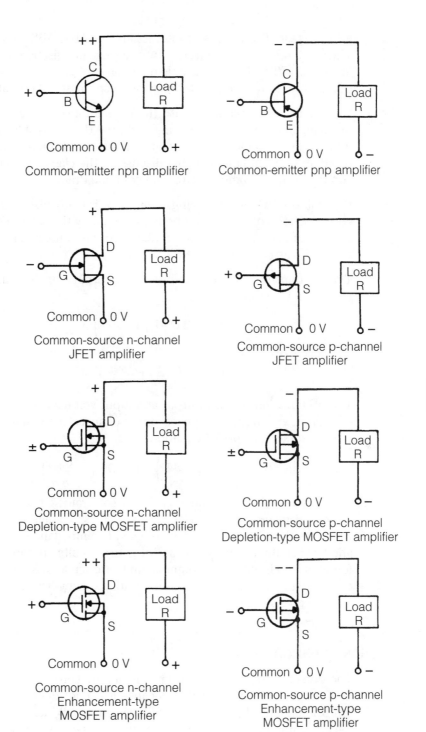

Common-emitter npn amplifier

Common-emitter pnp amplifier

Common-source n-channel
JFET amplifier

Common-source p-channel
JFET amplifier

Common-source n-channel
Depletion-type MOSFET amplifier

Common-source p-channel
Depletion-type MOSFET amplifier

Common-source n-channel
Enhancement-type
MOSFET amplifier

Common-source p-channel
Enhancement-type
MOSFET amplifier

55

■ **3-6** *You MUST know the polarities of voltages on three-terminal am-plifying devices.*

Three-terminal and four-terminal components and basic circuits

into the test. The classes of interest are: A, AB1, AB2, B, and C. There are other classes but they are not usually discussed or utilized in a CET test. A few sample questions are given at the end of this chapter to test your knowledge of these classes of operation.

You should also be familiar with the configurations of amplifiers. Speaking in terms of bipolar transistors, the three possible configurations are: common emitter, common base, and common collector. If an FET amplifier is being discussed, the classifications are: common source, common gate, and common drain.

Regardless of which amplifying component is being discussed, the characteristics of these amplifiers are generally the same. For example, common-base bipolar circuits have the same general characteristics as common-gate FET amplifiers.

The relationship between the input and output signals of an amplifier can be expressed in dB:

$$dB = 10 \: LOG_{10} \left(\frac{P_{OUT}}{P_{IN}} \right)$$

$$dB = 20 \: LOG_{10} \left(\frac{V_{OUT}}{V_{IN}} \right)$$

The dB equation relating output and input voltage is only valid if the output and input resistances, or impedances, are equal. If they are not equal, the correct equation is:

$$dB = 20 \: LOG_{10} \left(\frac{V_{OUT}}{V_{IN}} \right) + 10 \: LOG_{10} \left(\frac{R_{IN}}{R_{OUT}} \right)$$

As a general rule the common collector (or drain) circuits, which are called followers, have a power gain. Their voltage gain is always less than 1.0 (one). They are used for impedance matching and level shifting. Common base (or gate) circuits are most often found in RF circuits. The common emitter circuit (or source) circuits are the most often-used configurations. They have both voltage and power gain.

Practice question

The detector stage of a receiver is generally considered to be a high-output impedance circuit because it cannot deliver any significant amount of power to the following stage. Which of the following amplifier configurations would be best to match the high-output impedance detector stage to a low-input impedance amplifier?

A. Common emitter
B. Common base
C. Common collector

Answer: Choice C. This circuit, which is also known as an *emitter follower* or *source follower*, has a relatively high input impedance and a lower output impedance.

While on this subject, some special circuits are used to increase the input impedance of bipolar transistor amplifiers. One of the most important of these is the bootstrap circuit (Fig. 3-7). When the input signal at A goes positive in this circuit, the voltage at the top of the emitter resistor (the positive-going voltage) is coupled to B by capacitor C. So, the voltage at both sides of base resistor R go positive at the same time. The overall result is that the change in signal voltage across the input resistance does not increase the current through R. If you increase the voltage across a resistor and it does not increase the current very much the resistance must be very large. That is the basic concept of the bootstrap circuit.

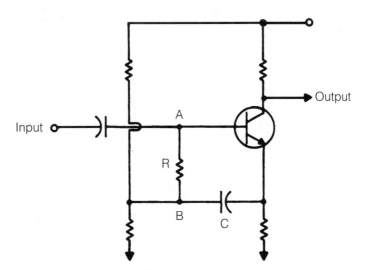

■ **3-7** *An example of a bootstrap circuit.*

Note: The word *bootstrap* is also used for other types of circuits. For example, one kind of bootstrap circuit is used to linearize sweep circuits. You will not be confused between the types of bootstrap circuits in the CET test because they will be clearly identified.

Three-terminal and four-terminal components and basic circuits

You must be thoroughly familiar with the various methods used to bias the amplifying devices. The bias circuits for bipolar transistors are shown in Fig. 3-8.

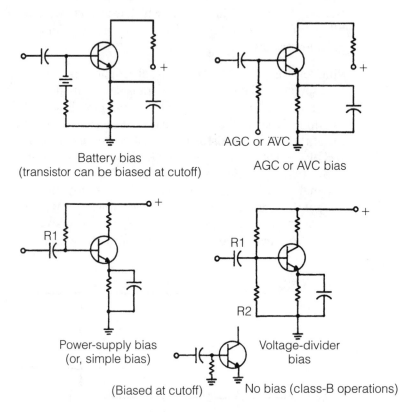

Battery bias
(transistor can be biased at cutoff)

AGC or AVC bias

Power-supply bias
(or, simple bias)

(Biased at cutoff)

Voltage-divider
bias

No bias (class-B operations)

■ **3-8** *Methods of biasing bipolar transistors.*

Not all of these bias circuits have equivalents in field-effect transistor or tube circuitry. For example, MOSFET or JFET amplifiers can be biased by putting a resistor in the source circuit. This is not possible with a bipolar transistor amplifier. However, you will find emitter resistors in those circuits. The purpose of those resistors is to stabilize the transistor amplifier against changes in temperature. You are likely to be asked about this emitter resistor. Be sure that you do not ascribe biasing to its purpose.

Figure 3-9 shows some special amplifier configurations that you should understand. They are not covered in great detail here, but they are identified in the following paragraphs.

The Darlington amplifier is sometimes called a beta-squared amplifier. The reason is that, with this combination, the beta of the

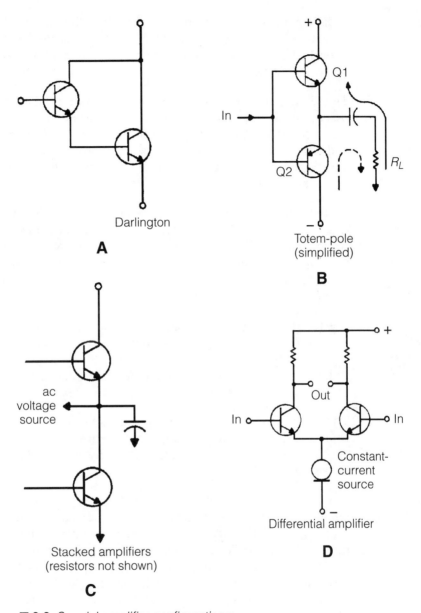

Darlington

A

Totem-pole
(simplified)

B

ac
voltage
source

Stacked amplifiers
(resistors not shown)

C

Out

In

In

Constant-
current
source

Differential amplifier

D

■ **3-9** *Special amplifier configurations.*

combined transistors is equal to the product of the beta of each transistor. Because Darlingtons are very often connected inside the same case and they are selected to be very closely matched, their betas are the same. Therefore, the beta of the configuration is beta X beta or beta squared.

A disadvantage of Darlington circuits is their high internal power dissipation. Their best advantage is that it is possible to get a relatively high gain and input resistance compared to other power-amplifier configurations.

The totem-pole circuit is popular for transistor output circuitry. In this particular example, *long-tailed bias* is used. The term means that a positive supply voltage is used at one end of the amplifier and a negative supply voltage is used at the other end. Long-tail bias permits the load to be operated at or near ground potential when the circuit is idling.

On one-half cycle, Q1 conducts (as shown by the solid arrow). On the next half cycle, Q2 conducts (as shown by the broken arrow). Notice that the direction of conduction through R_L is opposite on the two half cycles. That means that an ac current is flowing through the load resistor. In practice, the load resistor for the totem pole might be a speaker.

The totem-pole amplifiers (also known as *complementary amplifiers*) do not have equivalents in vacuum-tube circuitry. This is a circuit combination that permits 180 degree out-of-phase signals to be obtained from a single input. A center-tapped transformer can be used for the same purpose.

The stacked amplifiers give technicians considerable trouble because the dc circuits for the amplifiers are in series. You can think of the two amplifiers as being a dc voltage divider in addition to being signal amplifiers.

You might be asked a question on the CET test about how differential amplifiers or operational amplifiers reject common-mode signals. You might be asked about reflex circuitry in the Journeyman Consumer CET test. Remember that an amplifier can be made to amplify two different frequencies at the same time. For example, in Fig. 3-10, the IF amplifier also serves as the audio voltage amplifier. The advantage of this reflex circuit is that one amplifier can be used for two different signals, so the circuit is less expensive than using two amplifiers.

Noise in amplifiers

You should be able to identify some types of noise in amplifiers. *Partition noise* occurs when a charge carrier moves from the

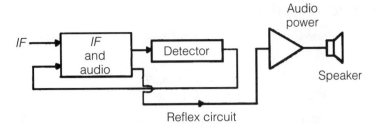

■ **3-10** *An example of a reflex circuit.*

emitter to the collector of a bipolar transistor. It has two different choices of a direction to go. For example, as an electron moves from the emitter to the collector in an NPN transistor, it reaches a point where it could go to either the base or to the collector. Both of those electrodes are positive. At any one instant, there will be a slight difference between the number of charge carriers that go to the base and the number that go to the collector.

The random selection of paths produces very small changes in the collector current. These small changes, when flowing through the collector load resistor, will produce the undesirable noise signals called *partition noise*. Partition noise is a problem in bipolar transistors but not in FETs.

MOSFETs have virtually no partition noise because the charge carriers move from the source to the drain without any alternate paths. Therefore, MOSFETs are strongly preferred for high-frequency applications, such as RF amplifiers and mixers.

Any time that current flows through a resistor, a noise signal will be present as a result of thermal agitation. Thermal agitation noise (also called *Johnson noise*) is especially prevalent in bipolar and MOSFET devices. In the MOSFET devices, the channel between the source and the drain is a resistive material.

Thermal agitation noise increases with an increase in temperature. It also increases with an increase in bandwidth and resistance values.

The equation for noise voltage shows how temperature (T), bandwidth (B), and resistance (R) affect noise voltage (V_n).

$$V_n^2 = 4kTBR$$

where: k is Boltzmann's constant
T is the temperature in degrees Kelvin
B is the bandwidth in Hertz
R is the resistance in Ohms

Practice question

Increasing the temperature of a semiconductor amplifier will cause:

A. an increase in amplifier noise.
B. a decrease in amplifier noise.

Answer: A. Thermal agitation noise is a problem in RF amplifiers and mixers.

Distortion in amplifiers

If you connect two different signal frequencies across a linear resistor, the signals do not combine in any way. All you end up with is the two individual frequencies. However, if you take the same two frequencies and introduce them to a nonlinear device one of the signals will modulate the other. That explains why converters, mixers and detectors are nonlinear circuits.

In a class-A amplifier no modulation between frequencies should occur if the amplifier is perfectly linear. Unfortunately, there is no such thing as a perfectly linear amplifier. Therefore, there will always be some cross modulation between signals—even in class-A amplifiers. This cross modulation is known as *intermodulation distortion*. It is a form of nonlinear distortion. The amount of intermodulation distortion can be minimized by operating the amplifier in the most linear portion of its characteristic curve.

If the amplifying device is not biased in the center of its linear characteristic, it is possible for a high-amplitude input signal to be clipped at one end or the other. In other words, either the positive or negative peaks will be clipped. This clipping has the same effect as introducing a great amount of distortion into the signal.

You will remember from your basic theory that a sine wave has absolutely no harmonic content. If you apply a pure sine wave to an amplifier in which clipping occurs, a large number of harmonics will be in the output. For that reason, *clipping* is a form of harmonic distortion.

Sometimes an amplifier is purposely overdriven so that both the positive and negative peaks are clipped. The result is a square-wave output signal that has many odd harmonics. Frequency multipliers are sometimes made this way.

No amplifier can produce the same amount of gain to all frequencies in its bandwidth. Any change in the gain at different frequencies is a form of frequency distortion.

The passive components in the output of the amplifier tend to shift the phase of signals at one frequency more than another. That is known as *phase-shift distortion*.

Special ratings

Anything that you do to increase the gain of an amplifier will automatically decrease its bandwidth! Conversely, anything that you do to increase its bandwidth will decrease its gain! Many technicians have corresponded with us about this point. They somehow believe that it is possible to increase both the gain and the bandwidth, but so far, no one has been able to show us how to do it.

An important amplifier rating is the gain-bandwidth product. It is a constant obtained by multiplying the gain by the bandwidth. Obviously, a transistor or FET with a higher gain-bandwidth product can produce a greater gain and greater bandwidth, but there is still a tradeoff within the component for the two measurements.

A gain-bandwidth product is a frequency. It is the frequency at which the beta drops to a value of 1.0. The gain-bandwidth product is important for two reasons: it shows that there is a tradeoff between these two parameters and it is a rough indication of the high-frequency capability of a given transistor or amplifying device.

The bandwidth of an amplifier is often defined as the range of signals between the half-power points on its frequency-response power. Another way of saying this is: it is the range of frequencies between the points where the power is down 3 dB.

For a response curve that shows voltage (or current) versus frequency (instead of power versus frequency), the bandwidth will be between the points where the signal-voltage (or current) amplitude is down 6 dB.

Remember that a graph showing frequency versus power or frequency versus voltage (or current) is called a *bode plot*. Frequency versus phase shift is sometimes shown on the same bode plot.

The alpha and the beta of a transistor drop off as the frequency increases. Two terms you might encounter are *alpha cutoff* and *beta cutoff*. Those are frequencies at which the alpha and beta drop to 70.7 percent of their values at 1 kHz.

63

The voltage gain of an amplifier is simply the output signal voltage divided by the input signal voltage. For some reason, technicians want to complicate this simple relationship, so they might miss a very simple question on the voltage gain of a given amplifier.

Likewise, the power gain is the output power divided by the input power. This refers to signal power. It has nothing to do with the total input power of the amplifier. The total input power would include the operating power that is necessary as a result of the application of dc operating voltages and currents.

Thyristors

Thyristors are electronic semiconductor devices that remain in the OFF condition until triggered by an input current or voltage. Once triggered they snap into the ON condition. Thyristors can be two-terminal, three-terminal, or four-terminal electronic components. Most are made with doped silicon, however, doped gallium-arsenide thyristors have also been manufactured. Figure 3-11 shows some popular thyristors in alphabetical order.

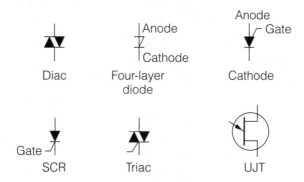

■ **3-11** *Popular thyristors shown in alphabetical order.*

Practice questions

1. The circuit in Fig. 3-12 is a:
    A. push-pull amplifier.
    B. direct-coupled complementary amplifier.

2. Which of the following is correct for the circuit in Fig. 3-13?
    A. If the switch is momentarily turned on then off, the lamp will go on and then off and stay off.
    B. If the switch is momentarily turned on then off the lamp will go on and stay on.

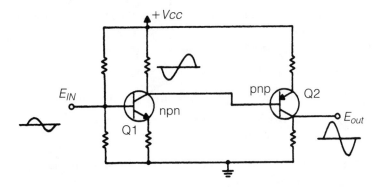

■ 3-12

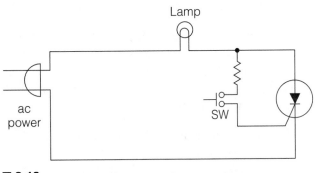

■ 3-13

3. Which of the following statements is correct?

   A. A triac acts like a back-to-back connection of SCRs.

   B. A UJT acts like a back-to-back connection of SCRs.

4. What is the time constant of a 100-millihenry coil connected in series with a 100-kilohm resistor?

   A. 0.1 second

   B. 1.0 second

   C. Neither choice is correct.

5. In order for modulation to take place two sine-wave signals with different frequencies must be applied to a:

   A. linear device.

   B. nonlinear device.

6. Anything you do to increase the gain of an amplifier will automatically decrease:

   A. its operating frequency.

   B. its bandwidth.

Three-terminal and four-terminal components and basic circuits

7. What type of amplifier is sometimes called a *beta-squared amplifier*?

   A. Darlington
   B. Push-pull

8. You would expect very little partition noise in an:

   A. N-channel MOSFET.
   B. NPN transistor.

9. A transistor that does not have a collector is a:

   A. unijunction transistor.
   B. junction transistor.
   C. MADT.
   D. PEP transistor.

10. A transistor that can be operated in the enhancement mode is a:

    A. MOSFET.
    B. MADT.
    C. PEP.
    D. drift transistor.

11. Another name for a three-layer diode is:

    A. tunnel diode.
    B. Shockley diode.
    C. diac.
    D. LDR.

12. If a small amount of negative voltage is in the gate of an SCR, it will:

    A. decrease the leakage current.
    B. prolong the life of the SCR.
    C. decrease the breakover voltage.
    D. increase the breakover voltage.

13. If measurements show that electron current is flowing from the anode to the cathode in a zener diode, the indication is:

    A. that the diode is in the circuit backwards.
    B. that the direction of current flow is proper.

14. Which of the following is an application of a tunnel diode?

    A. Detector
    B. Oscillator
    C. Voltage regulator
    D. High-voltage rectifier

15. Increasing the reverse bias on a varactor diode will:

    A. decrease its capacitance.
    B. increase its capacitance.
    C. not affect its capacitance.

16. A reverse bias on a pn-junction diode will always destroy a germanium diode if the:

    A. reverse bias reaches the zener value.
    B. reverse bias is pure dc.
    C. reverse bias is pulsed.

17. Which diode has a negative resistance section on its characteristic curve?

    A. Diac
    B. Breakdown diode
    C. Shockley diode
    D. Esaki diode (also known as a *tunnel diode*)

18. The base-collector junction of a transistor is normally:

    A. always some form of hot-carrier construction.
    B. reverse biased.
    C. forward biased.
    D. always passivated.

19. Current gain in a common-emitter transistor circuit is called the:

    A. alpha.
    B. impedance transfer.
    C. gain-bandwidth product.
    D. beta.

20. A ratio-detector circuit might use:

    A. four-layer diodes.
    B. three-layer diodes.
    C. varactor diodes.
    D. pn junction diodes.

21. A class-A amplifier might be made with:

    A. a triac.
    B. a UJT.
    C. a MOSFET.
    D. an SCR.

22. The electrode on a CRT that corresponds to the source of JFET is the:

    A. plate.
    B. control grid.

C. cathode.

D. screen.

23. Which of the following conducts in either of two directions?

A. NPN transistors

B. Triacs

C. SCRs

D. PNP transistors

24. The polarity of voltage on the gate, with respect to the drain of an N-channel JFET is normally:

A. positive.

B. negative.

25. The frequency at which the common-emitter current gain drops to 70.7 percent of its value at 1 kHz is called:

A. alpha cutoff frequency.

B. admittance transfer frequency.

C. gain-bandwidth product.

D. beta cutoff frequency.

26. Which of the following is nearest to a thyratron in its operation?

A. MOSFET

B. SCR

C. MADT

D. Diac

27. Which of the following is a type of thyristor?

A. Tunnel diode

B. LDR

C. Junction diode

D. Diac

28. Which of the following types of diodes is sometimes used as a fast-acting switch?

A. Diacs

B. Junction diodes

C. Tunnel diodes

D. Zener diodes

29. A P-channel FET should have a gate/cutoff voltage that is:

A. positive, with respect to the voltage on the source.

B. negative, with respect to the voltage on the source.

30. Which of the following is not a thyristor?

A. An Esaki diode

B. An SCR

C. A triac

D. A diac

31. The voltage on the collector of an NPN transistor audio amplifier is positive, and the voltage on the base is negative, with respect to the emitter. Which of the following is true?

    A. The polarity of the voltage on the collector is wrong.

    B. The polarity of the voltage on the base is wrong.

    C. The polarities of both voltages are wrong.

    D. This is normal.

32. Another name used for a four-layer diode is:

    A. a diac.

    B. an Esaki diode.

    C. a hot-carrier diode.

    D. a Shockley diode.

33. Which of the following is an example of a breakover diode?

    A. LDR

    B. PN junction diode

    C. Shockley diode

    D. Tunnel diode

34. The operation of a diac is most nearly like the action of a:

    A. neon lamp.

    B. JFET.

    C. thyratron.

    D. phanatron.

35. To forward bias a rectifier diode:

    A. its anode is made positive, with respect to the cathode.

    B. its anode is made negative, with respect to its cathode.

36. Which of the following semiconductor devices would be used to obtain a relatively constant voltage?

    A. Four-layer diode

    B. Diac

    C. Zener diode

    D. Tunnel diode

37. The minority charge carriers in P-type material are:

    A. positrons.

    B. neutrons.

    C. electrons.

    D. holes.

38. Another name for Esaki diode is:

    A. tunnel diode.
    B. Shockley diode.
    C. diac.
    D. breakdown diode.

39. Charge carriers do not exist in a diode's:

    A. barren straits.
    B. no-man's land.
    C. depletion region.
    D. waste region.

40. An SCR electrode that corresponds to the grid of its equivalent tube device is the:

    A. gate.
    B. drain.
    C. emitter.
    D. anode.

41. In a common-base transistor circuit, the current gain is called the:

    A. alpha.
    B. impedance transfer.
    C. beta.
    D. gain-bandwidth product.

42. The common-base current gain drops to 70.7 percent of its 1-kHz value at a point called:

    A. gain-bandwidth product.
    B. admittance transfer frequency.
    C. beta cutoff frequency.
    D. alpha cutoff frequency.

43. The common-emitter forward-current transfer ratio is unity at a point called:

    A. beta cutoff frequency.
    B. alpha cutoff frequency.
    C. gain-bandwidth product.
    D. none of these choices is correct.

44. The electrode on a JFET that corresponds to the plate of a triode is called:

    A. drain.
    B. gate.
    C. base.
    D. source.

45. Which of the following is a bipolar transistor?

    A. NPN
    B. Diac
    C. MOSFET
    D. JFET

46. Which of the following would be grounded in a source-follower?

    A. Drain
    B. Collector
    C. Gate
    D. Source

47. To stop an SCR from conducting:

    A. drive the gate voltage negative with a pulse.
    B. make the anode and cathode voltages equal.

48. When a UJT is used as a relaxation oscillator, the timing capacitor is connected between:

    A. emitter and base.
    B. emitter and collector.
    C. a base and source.
    D. None of these choices is correct.

49. As a general rule:

    A. triodes are noisier than pentodes.
    B. tube noise is greater than FET noise.

50. For the circuit in Fig. 3-14 the resonant frequency:

    A. can be adjusted with R_C or R_L.
    B. cannot be adjusted by R_C or R_L.

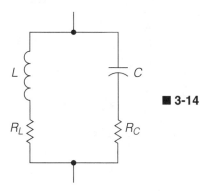

■ 3-14

51. If the collector current of a transistor flows all of the time, the transistor is being operated in:

A. class A.
B. class B.
C. class C.

52. Which of the following types of bias is not used with transistors?
    A. AGC bias
    B. Power-supply bias
    C. Battery bias
    D. Self-bias using an emitter resistor

53. The base voltage of an NPN transistor is:
    A. positive, with respect to the collector voltage.
    B. negative, with respect to the collector voltage.

54. Which is the most efficient class of amplifier?
    A. Class A
    B. Class AB2
    C. Class B
    D. Class C

55. Transistors are seldom operated in:
    A. class A.
    B. class AB.
    C. class B.
    D. class C.

Answers

Question	Answer
1.	B
2.	B
3.	A
4.	C
5.	B
6.	B
7.	A
8.	A
9.	A
10.	A
11.	C
12.	D
13.	B
14.	B
15.	A

16.	A
17.	D
18.	B
19.	D
20.	D
21.	C
22.	C
23.	B
24.	B
25.	D
26.	B
27.	D
28.	C
29.	A
30.	A
31.	B
32.	D
33.	C
34.	A
35.	A
36.	C
37.	C
38.	A
39.	C
40.	A
41.	A
42.	D
43.	A
44.	A
45.	A
46.	A
47.	B
48.	A
49.	B
50.	A
51.	A
52.	D
53.	B
54.	D
55.	D

Analog circuits

4

Associate level *This material is related to Sections I through V in the Associate-level CET test.*

Journeyman level *This material is related to Section XI of the Consumer CET test. It is also related to other Journeyman tests.*

Here is a definition of an *analog circuit*: it has an output signal at all times when there is an input signal. Also, its output signal amplitude is continuously variable when the input voltage or current is varied from its minimum value to its maximum value.

By contrast, a digital circuit normally operates in either of two fixed voltage values. It has no stable condition between those two values.

75

Practice question

A class-C amplifier has an output signal only when there is a positive input signal. Is it an example of a digital circuit?

Answer: No. It does not have only two output voltage levels.

Note: Before you take any CET test, review the characteristics of class-A, class-B, and class-C amplifiers. Also, review class-AB1 and class-AB2 amplifiers.

Amplifiers

Amplifier coupling circuits

As a technician, you should be thoroughly familiar with the methods of coupling a signal from one amplifier to another. When the output signal of one amplifier feeds to the input of another amplifier, the combination is said to be *cascaded*. Do not confuse this term with *cascoded*, which is a special type of low-noise RF amplifier. Figure 4-1 reviews the methods of coupling amplifiers.

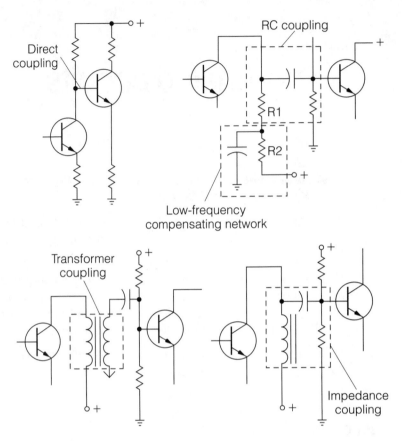

■ 4-1 *Methods of coupling amplifiers.*

When the output of one amplifier is connected to the input of the next amplifier with a piece of wire, the two amplifiers are said to be *direct coupled*. In the days of vacuum tubes, the direct-coupled circuits were called *Loftin-White amplifiers*.

The obvious advantage of direct coupling is that it has a very wide frequency response. That is achieved because no reactive component is in the coupling circuit. However, reactance is at the output of the one amplifier and at the input of the other amplifier. It is usually capacitive to common and it limits the high-frequency response of the direct-coupling circuit.

Resistor-capacitor coupling (which is often called *RC coupling*) is the least expensive method of coupling two amplifiers. You might think that direct coupling would be simpler and cheaper, but that method has a problem called *level shifting*. Usually, the output of one amplifier is at a different dc level than the input of the other, so the second amplifier must usually be operated at a higher voltage. Unless an interfacing circuit is used, the power supply must be more elaborately designed. This is not only true because of the different dc levels, but if any ripple is on the power supply output, that ripple will be delivered to the second amplifier and will be treated as an ordinary signal. In other words, the ripple will be amplified considerably.

RC coupling has a limited low-frequency response because the coupling capacitor has a high reactance to low frequencies. To circumvent the problem of poor low-frequency response, some amplifiers use a *low-frequency compensating network* (Fig. 4-2).

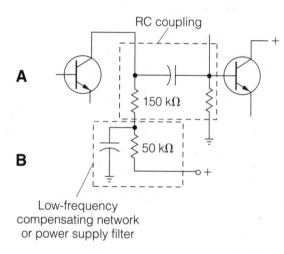

■ **4-2** *Low-frequency compensating network.*

At low frequencies, the capacitance at the junction of the two resistors in the collector circuit is nearly an open circuit. In other words, it has a high reactance. Therefore, at low frequencies, the gain of the amplifier is set by a high value of collector load resistance, which consists of the two resistors in series.

Even though the low-frequency signal is amplified with more gain, the actual signal that arrives at the next amplifier is still not greatly increased in amplitude because of the increased reactance (opposition) of the coupling capacitor. Thus, the gain is increased to overcome the high loss in the coupling capacitor at low frequencies.

At high frequencies, the coupling capacitor becomes a short circuit (virtually), so the signal is coupled from one amplifier to the next with very little loss. At the same time, the capacitor at the junction of the collector resistors also becomes a short circuit for the signal. For all practical purposes, the lower resistor (R2) is shorted out, and the collector load resistance is reduced to one resistor. The reduced collector resistance reduces the gain of the stage, but less gain is needed because the loss from the coupling capacitor is greatly reduced.

The low-frequency compensating network looks identical to a decoupling filter for a power-supply line. You can tell which it is by the value of the resistance in the circuit, and also by the value of the capacitor. Specifically, the resistor must be relatively large if it's a low-frequency compensating network. Otherwise, it would not materially add to the collector load resistance when the low frequencies are being boosted.

On the other hand, if it's a decoupling filter, the resistance will be very low (10 to 100 ohms is common). The reason is that it is not desired to drop any appreciable amount of dc power-supply voltage across the decoupling filter.

Another kind of coupling is obtained with a transformer. Transformer coupling has an important advantage: It can be tuned to select some particular frequency or band of frequencies. That is especially true when the transformer between stages is an air core or ferrite type. That situation is important to a technician. By looking at the transformer coupling between two stages, a technician can determine whether the circuit is designed for audio (using an untuned iron-core transformer) or for RF (using a tuned air-core transformer).

Transformer coupling also has the advantage of being broadband, but has the disadvantage of being too frequency selective in broadband circuits. Don't forget that transformer coupling is useful in phase-splitting applications. When a phase inverter is needed for push-pull operation, a transformer can easily supply the 180-degree out-of-phase signals. Also, the transformer can be used for impedance matching.

Impedance coupling between amplifiers is obtained with combinations of resistance, inductance, and capacitance. Impedance coupling might use a tunable component so that it can pass a specific band of frequencies. You will see that type of circuit used in IF amplifiers in television systems. They replace the transformer coupling, which has the disadvantage that it is more difficult to obtain a complete 6-MHz bandwidth over the complete range of IF frequencies.

Practice question

By way of review, mark the polarity (+ or −) of voltage on each of the amplifying devices in Table 4-1. Observe that the current input electrode (emitter or source) is marked 0 V. When measuring voltages on an amplifying device, the usual procedure is to compare the voltages with the voltage on the current input electrode. In other words, that electrode is considered to be 0 V.

Answer: After you have marked the polarities, compare your answers with Fig. 3-6. You must know these polarities when you are troubleshooting at the component level.

Characteristics of amplifier configurations are shown in Table 4-2, and some basic equations for amplifying devices are shown in Table 4-3.

Practice question

Is the circuit enclosed with broken lines in Fig. 4-2A:

A. low-frequency compensating network?
B. power supply filter?

Answer: Choice B is correct. Observe the high value of resistance.

Integrated circuit (IC) operational amplifiers (op amps)

An operator in mathematics is a mathematical symbol (such as +, −, ÷, and ×) that tells the person working the problem what is supposed to be done. For example, the plus sign tells the person to add the numbers.

An IC operational amplifier is a very-high gain amplifier that is constructed in an integrated circuit package. It can perform arithmetic operations and many other tasks.

It can be proved with a mathematical analysis that any time a high-gain amplifier utilizes negative feedback, the output of that amplifier can be made dependent only upon its feedback network!

There are several requirements for IC operational amplifiers. The ideal IC operational amplifier must have a very high open-loop gain. (The open-loop gain of an amplifier is its gain without feedback.) It must have a high input impedance and a low output impedance.

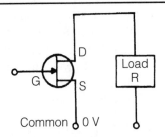

Common-source n-channel
JFET amplifier

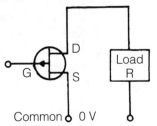

Common-source p-channel
JFET amplifier

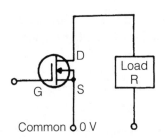

Common-source n-channel
Depletion-type MOSFET amplifier

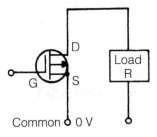

Common-source p-channel
Depletion-type MOSFET amplifier

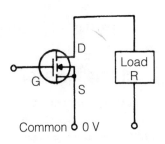

Common-source n-channel
Enhancement-type
MOSFET amplifier

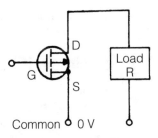

Common-source p-channel
Enhancement-type
MOSFET amplifier

80

■ Table 4-2 Characteristics of amplifier configurations.

Common electrode	Input signal goes to	Output signal comes from	Characteristics
Plate, collector, or drain	Control grid, base, or gate	Cathode, emitter, or source	Circuit is called a follower. Voltage gain less than 1.0. May have power gain. Matches high impedance to low impedance.
Control grid, base, or gate	Cathode, emitter, or source	Plate, collector, or drain	Common-connected control electrode acts as a Faraday shield between the input and output signal connections. Good for high-frequency operation. Matches low impedance to high impedance.
Cathode, emitter, or source	Control grid, base, or gate	Plate, collector, or drain	Most common connection. Best compromise for both voltage and power gain. Input and output impedances are reasonably high.

■ Table 4-3 Equations for amplifying devices.

Bipolar transistors are also called BJTs (for Bipolar Junction Transistors)

$$\text{dc alpha } (\alpha_{DC}) = h_{FB} = \frac{I_C - I_{CBO}}{I_E} = \frac{I_C}{I_E}$$

$$\text{dc beta } (\beta_{DC} = h_{FE} = \frac{I_c - I_{CBO}}{I_B} = \frac{I_C}{I_B}$$

$$\text{ac alpha } (\alpha_{ac}). = h_{fb} = \frac{\Delta I_c}{\Delta I_e}$$

$$\text{ac beta } (\beta_{ac}) = h_{fe} = \frac{\Delta I_c}{\Delta I_b}$$

alpha is always less than 1.0
beta is always greater than 1.0

alpha cutoff frequency   The point on a graph where alpha always drops to 70.7% of maximum.

Beta cutoff frequency   The point on a graph where beta drops to 70.7% of maximum

$$\alpha = \frac{\beta}{1 + \beta}$$

$$\beta = \frac{\alpha}{1 - \alpha}$$

■ **Table 4-3 Continued.**

MOSFETS and JFETS

Under normal operating conditions the gate current = I_G = 0.

$$g_m \text{ (transconductance)} = \frac{I_D}{I_{GS}}$$

In the depletion mode, the gate bias equals the drain current (I_D) multiplied by the resistance of the source resistor (R_s)

In the enhancement mode, there is no drain current unless the drain is forward biased. The drain is forward biased with a voltage divider.

Figure 4-3 shows a bode plot of a 741 operational amplifier. The 741 op amp has become a standard (or, "benchmark") by which other operational amplifiers can be compared.

Basic op amps

A *bode plot* is also called the *response curve*. It is an excellent way of showing the tradeoff between gain and bandwidth. Two examples are marked with broken lines on the bode plot of Fig. 4-3. The first example shows that when the gain is 10 the bandwidth is 100 kHz. However, when the gain is increased to 100, as shown by the second example, the bandwidth decreases to 10 kHz. Therefore, increasing the gain in the op amp has reduced the bandwidth. That is a tradeoff for all types of amplifiers.

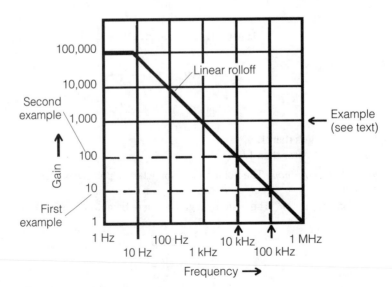

■ **4-3** *Bode plot of a #741 op amp.*

Practice question

An operational amplifier is being used in a system that requires a bandwidth of 100 Hz. The bode plot of that op amp is shown in Fig. 4-3. What is the gain of this operational amplifier?

A. $A_v = 100$
B. $A_v = 1000$
C. $A_v = 10,000$

Answer: Choice C is correct. When the operational amplifier has a voltage gain of 10,000 ($A_v = 10,000$), the bandwidth is 100 Hz.

Here is the equation for gain-bandwidth product:

$$Gain\text{-}bandwidth\ product = Gain \times Bandwidth$$

The gain-bandwidth product is a frequency measurement. Because the gain-bandwidth product is a constant, it follows that any increase in gain will automatically result in a decrease in bandwidth. Also, any increase in bandwidth will automatically result in a decrease in gain.

The lesson to be learned here is that you cannot increase one without decreasing the other.

Practice question

Look at the amplifier circuit in Fig. 4-4. The purpose of the emitter resistor is to stabilize the amplifier against temperature changes. (It is NOT put in the circuit to bias the amplifier!) The capacitor in parallel with R_E is used to prevent *degeneration* (loss of gain caused by the presence of the resistor).

Why does the designer sometimes omit the capacitor in the circuit?

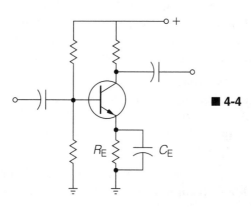

■ 4-4

Answer: When the capacitor is omitted the gain of the amplifier is reduced. However, the designer might tolerate the loss of gain in order to get a wider bandwidth. The wider bandwidth means the frequency response is better.

Although all amplifiers have a gain-bandwidth tradeoff, there is no predictable tradeoff between those parameters unless there is a linear rolloff, as shown by the bode plot in Fig. 4-3.

Advertisements from the mid-1940s brag about vacuum-tube operational amplifiers with open-loop voltage gains as high as 5000. The IC operational amplifiers used today have open-loop voltage gains in the range of 100,000 to well over 3 million!

Another requirement of an ideal operational amplifier is that it must have a differential input. It means that a differential amplifier is used at the input stage.

Figure 4-5A shows a simplified circuit for a differential amplifier. It is assumed when you take the CET test that you understand how a differential amplifier works.

A common-mode operation exists when you connect the two inputs (A and B) together. That way the input signal is simultaneously applied to both sides of the differential amplifier. Theoretically, if the differential amplifier is perfectly balanced and a signal is applied to the two inputs simultaneously, the output should be zero volts. That will be true only if it is a perfect differential amplifier. If any unbalance exists in the two differential amplifier sections a common-mode input will produce an output signal. In the real world, no differential amplifier is perfect.

Many circuits are available for setting the balance between the two input sections of the differential amplifier in an op amp. To review its operation, suppose one of the amplifiers in Fig. 4-5A is conducting harder than the other in the absence of any signal. That would mean the amplifier is unbalanced and a dc output voltage could be measured across the output terminals. Such an unbalance would be highly undesirable because there would always be an output voltage even though there would be no input voltage.

To eliminate that possibility, a dc bias can be applied to one or both of the two amplifiers. The bias voltage is adjusted so that the output of the two amplifiers is identical. With that offset bias adjustment, the output of the operational amplifier can be made very nearly equal to 0 V when there is a common-mode input.

A second method of balancing differential amplifiers is shown in Fig. 4-5B. The arm of the variable resistor can be adjusted to balance out differences in branch currents. That eliminates the common-mode error voltage across the output terminals. Offset bias adjustments are sometimes made outside the IC package, as shown in Fig. 4-5C.

Figure 4-5A also shows how to use a FET to get a constant current flow into the differential amplifier. That is also a requirement for a differential amplifier. The FET connection is equivalent to a constant-current diode as shown in Fig. 4-5C. In fact, when you buy a constant-current diode, you actually get the FET connection.

A very high resistance is sometimes used as shown in Fig. 4-5B in non-IC differential amplifiers to simulate a constant-current device.

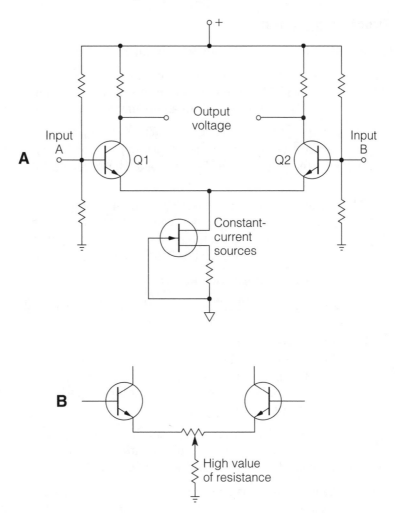

■ **4-5** *Why is the capacitor sometimes omitted?*

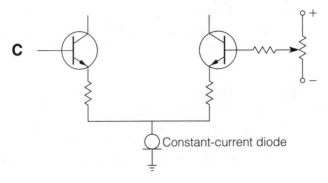

C

Constant-current diode

■ **4-5** *Continued.*

Practice question

How is a common-mode measurement made?

A. The input terminals of an op amp are shorted together and a signal is delivered to both inputs at the same time.
B. The output terminals of an op amp are connected to ground.

Answer: Choice A is correct. When the input terminals are connected together, and an ac signal is applied to that common connection, the output should be zero volts. That is called a *common-mode rejection test*.

A feature you would like to have in an operational amplifier is an enormously high open-loop gain and a very, very small common-mode output. When you buy an operational amplifier you can look in the specifications for the CMRR (common-mode rejection ratio).

The mathematical equation for CMRR is:

$$CMRR = \frac{Differential\ amplifier\ gain}{Common\text{-}mode\ gain} = \frac{A_{dm}}{A_{cm}} = \frac{(\beta)(R_E)}{h_{ie}}$$

For these equations:
β is the transistor beta
R_E is the emitter resistance
h_{ie} is the common-emitter transistor input resistance with the output signal short circuited.

A high CMRR is desirable. From the CMRR equation you can see that the desirable high CMRR occurs when transistors with a high β are used and the emitter resistance is high.

A high input impedance is needed so that the operational amplifier does not load the signal source. In many applications, the signal source is not able to deliver any appreciable amount of load current.

The output impedance should be nearly zero ohms so that it can be connected to a low-impedance device without having an undesirable mismatch. For a 741 operational amplifier, the output impedance is so low that when it is operating, you can short-circuit the output terminal to ground without having a disastrous overload current.

To achieve an output impedance that is very close to zero ohms, the output circuit of many operational amplifiers is in a *totem-pole configuration*. Figure 4-6 shows an example in simplified form. A speaker is shown in the output circuit of the amplifier. This circuit is also called a *complementary amplifier*.

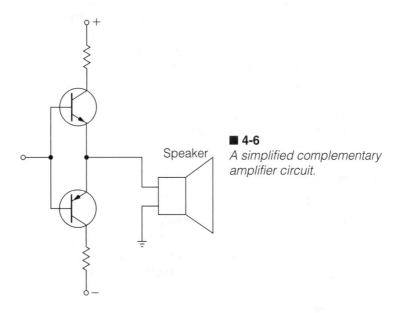

■ **4-6**
A simplified complementary amplifier circuit.

The circuit in Fig. 4-6 uses *long-tail bias*, which means that it uses both positive and negative power supplies. The output voltage is nearly half-way between the positive and negative voltages. In other words, the output is operating at nearly zero dc volts and nearly zero impedance.

Another important characteristic of an operational amplifier is its *slewing rate*. Slewing rate is a method of evaluating an operational amplifier for the speed with which it can change its output voltage when supplied with an input step voltage. In other words, if the input voltage goes from zero to some voltage value instantly, the output should (theoretically at least) also go to its maximum or minimum value instantly. No operational amplifier can actually achieve that, but the new IC op amps come very close.

Slewing rate is often measured in volts per microsecond. If you are going to use an operational amplifier in a high-frequency circuit or in a pulse circuit, you will certainly want one with a high slewing rate.

It is not uncommon to use operational amplifiers in digital circuits. For example, you can connect an operational amplifier to produce a pulse or a square-wave output. In order for the pulse to be applicable in a digital system, it must have a very fast rise time and very fast decay time. That can be achieved if the op amp has a high slewing rate.

Remember this important thing about digital circuits: The relaxation oscillators, counters, and many of the other circuits in a digital system operate only during the rise time or the decay time of an input pulse. If you have a long rise time or a long decay time (low slewing rate), you will not get a fast switching response. That is undesirable in digital systems.

An operational amplifier must have a rapid slewing rate if it is to reproduce a good-looking pulse or square wave in its output.

Op amp circuits

Having reviewed the important features of an op amp, it is now time to look at a few basic circuits. The following section is in no way intended to represent the total range of applications of operational amplifiers. A list of op amp applications would fill this book. The examples given in this chapter are typical. You might see related questions in a CET test.

The inverting amplifier

An inverting amplifier has an output signal that is 180° out of phase, with respect to the input signal. An op amp inverting amplifier connection is shown in Fig. 4-7. The gain of this circuit can be calculated by using the following equation:

$$A_\mathrm{v} = -\frac{R_\mathrm{f}}{R_\mathrm{i}}$$

where: A_V = amplifier voltage gain
R_f = the feedback resistance
R_i = the input resistance

The negative sign is used to indicate a phase inversion of 180°.

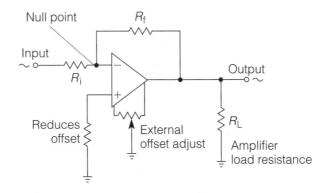

4-7

Practice question

The negative sign in the voltage gain equation for the circuit of Fig. 4-7 means:

A. the output signal is negative at all times with respect to common.
B. the output signal is 180° out of phase with the input signal.
C. it is the connection for the negative terminal of the power supply.
D. None of the choices is correct.

Answer: Choice B is correct.

Practice question

What is the gain of the op-amp circuit in Fig. 4-8?

Answer: $A = -\left(\dfrac{100,000}{4,000}\right) = -25$

Practice question

What is the gain of the op-amp circuit in Fig. 4-9?

Answer: The gain is –25. The gain depends only upon the feedback circuit. The amplifiers in Figs. 4-8 and 4-9 have the same feedback circuit.

The gain of the inverting op amp depends only on the feedback and input resistance circuitry. Remember, if the amplifier has a sufficiently high gain the op-amp gain equation will be accurate.

89

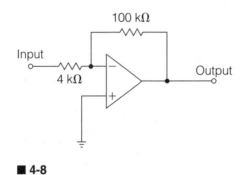

■ **4-8**

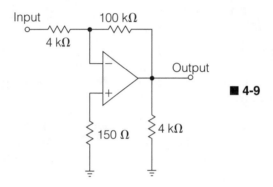

■ **4-9**

There is a good way to explain why the op amp output depends upon the resistance of R_f. The input resistance of the op-amp circuit must be relatively high. (Usually it is at least 1500 ohms.) That high input resistance prevents excessive loading by the input circuit.

The junction between the input resistance and the feedback resistance is at the inverting terminal. It is called the *summing point*. It is also called *virtual ground*.

If the op amp is operating properly, the voltage at that point should be nearly 0 V. The operational amplifier performs its job by maintaining the summing point at zero volts!

If the feedback resistance (R_f) is high—meaning a high R_f to R_i resistance ratio—the gain of the operational amplifier is also high. That is necessary so that when the amplifier output signal is dropped across the feedback resistor, the voltage will be zero at the summing point.

Conversely, if the feedback resistance is low—meaning a low resistance ratio—the gain of the operational amplifier does not have to be very high in order to produce enough signal through the feedback resistor to maintain the summing-point terminal at zero volts.

The noninverting amplifier

A symbol for the noninverting amplifier connection is shown in Fig. 4-10. Observe in this case that the feedback network is in place, and the gain of the amplifier still on that feedback network. That is necessary in order to achieve zero volts at the summing point. The gain of the noninverting op-amp circuit in Fig. 4-10 is given as:

$$A_v = 1 + \frac{R_f}{R_i}$$

This equation shows that the gain of the noninverting amplifier is slightly higher than for the inverting amplifier. However, if the ratio of the feedback resistance to the input resistance is high (meaning that the gain of the amplifier in a closed-loop configuration is high), the gain of the inverting and the noninverting amplifier is virtually the same.

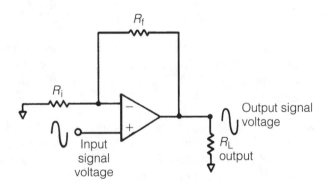

■ **4-10** *The op amp as a noninverting amplifier.*

The output signal is in phase with the input signal in this noninverting configuration. Another way of saying that is there is no phase inversion with this configuration.

Follower circuits (Fig. 4-11) have a high input impedance and a low output impedance. A second characteristic is that the follower does not invert the phase of the input signal, and, the gain of the circuit is slightly less than unity (1.0). These characteristics make it a simple matter to connect an operational amplifier as a follower.

This configuration is often referred to as a *buffer*. It is an amplifier that has a gain of about one, and, it is connected between two circuits to match impedance and/or to isolate one circuit from another.

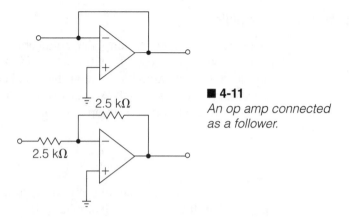

4-11
An op amp connected as a follower.

2.5 kΩ

2.5 kΩ

One of the circuits in Fig. 4-11 has no feedback resistance and no input resistance. However, the same gain of one can also be accomplished by using a feedback resistance. The second example in Fig. 4-11 shows the op amp with a feedback resistance of 2500 ohms and an input resistance of 2500 ohms.

In practice, the input resistor is selected to offer a high input impedance to the buffer. Its resistance might be selected to provide an impedance match. Manufacturers usually stipulate a minimum input resistance value.

A buffer made with resistors (shown in Fig 4-11) is not preferred. The gain of a noninverting amplifier is very slightly higher than the ratio of the feedback resistances, so the feedback circuit might have to be trimmed in order to get a gain of one.

Additional op-amp circuits

There are some additional applications of operational amplifiers that you should be familiar with. They are mentioned, but the individual circuitry is not covered. As an experienced technician, you should take time to review operational amplifier technology in depth because you might see applications in consumer products and on the Journeyman Consumer CET test.

Gyrator A gyrator is an operational amplifier circuit that is connected in such a way that it behaves like an inductor.

Comparator Because of its differential amplifier input, an operational amplifier is ideally suited for making a voltage comparison between two signals. For example, it can compare two dc voltages to determine which is more positive.

Timers

Many electronic timers are linear integrated circuits that can be used to delay the start of a signal or system. These very versatile timers are also utilized as oscillators, sensors, and other circuits. The 555 timer is one of the most popular IC timers. For its internal construction it uses two operational amplifiers in voltage comparator and shut-down circuitry. Figure 4-12 shows a 555 timer connected as a free-running (astable) oscillator.

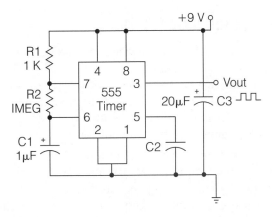

■ **4-12** *The 555 timer connected as an astable oscillator.*

Additional circuits you should know

Compander and expander The dynamic range of any signal is the range of values between its minimum and maximum amplitudes. If the dynamic range of a particular signal is greater than the dynamic range of the circuit to which it is being delivered, it can be passed through a compander. That circuit preserves the same relative voltages between the different parts of the signal but it decreases its total dynamic range. In contrast, an expander takes a relatively narrow range of signal voltages and increases its amplitudes to obtain a wide dynamic range.

Sample and hold Sample-and-hold circuits can be constructed with operational amplifiers. They are especially useful in analog-to-digital (A/D) converters and measurement circuits. At the request of an input gate signal, the sample-and-hold circuit maintains an output that is proportional to the amplitude of the input signal at the instant the gate signal appeared. The circuit output holds the voltage.

Active filter A passive filter uses only resistors, capacitors, and inductors. Active filters use amplifiers to obtain a desired output signal characteristic. Operational amplifiers are used extensively in active filter circuits. An example of an active filter made with an operational amplifier is shown in Fig. 4-13.

Multiplexers and demultiplexers A multiplexer selects one of a number of input signals and delivers it to a single output terminal. A demultiplexer delivers a single-input signal to a number of output terminals. Only one output terminal can be selected at any one time.

The phase-locked loop (PLL)

Some brands of early television receivers used phase-locked loop (PLL) circuits to maintain the horizontal oscillator on frequency. One example was the synchrolock type of automatic frequency control. That vacuum tube circuit uses a close-loop system that sets the oscillator frequency equal to the horizontal synchronizing pulses coming from the transmitter.

Figure 4-14 shows a block diagram of an integrated circuit PLL. An input signal frequency (f_{in}) is compared with the oscillator frequency (f_{osc}) of a voltage-controlled oscillator (VCO). If those frequencies are equal in frequency and phase there is no output correction voltage from the comparator.

Any difference in frequency or phase results in a comparator dc output correction voltage. That voltage passes through a low-pass filter. Its purpose is to ensure that the high-frequencies of the oscillator and incoming signals do not pass. The correction voltage must be a well-filtered dc voltage.

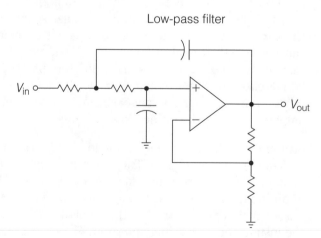

■ **4-13** *An example of an active filter.*

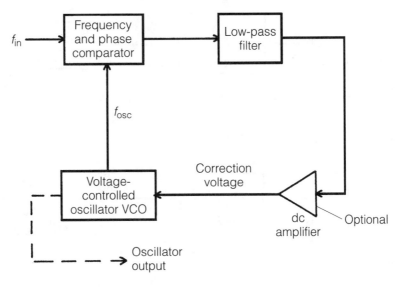

■ **4-14** *The block diagram of a PLL.*

An optional dc amplifier might follow the filter. It converts small changes in the dc correction voltage to large voltage changes. The VCO frequency is controlled by a varactor diode. If that frequency does not exactly equal the incoming frequency the correction voltage from the dc amplifier, or from the low-pass filter will adjust the varactor capacitance until the oscillator frequency and phase match the incoming signal.

One application of the circuit is to obtain an oscillator output frequency that is locked to the incoming frequency. The required output signal is delivered from the VCO as indicated by the broken arrow.

If the incoming signal is frequency modulated, the correction voltage will continuously try to correct the oscillator. The result is that the correction voltage will follow the modulating signal of the FM input.

Thus, the correction voltage is the same signal that modulated the FM signal. The overall result is that the PLL can be used as an FM detector. The audio output signal is taken from the output of the low-pass filter or the dc amplifier.

The phase-locked loop can also serve as a frequency synthesizer. One application is to obtain a local oscillator frequency that is always an exact multiple of the incoming frequency. Figure 4-15 shows principle of operation for this circuit. A programmable counter ($\div N$) has been added to the PLL circuit. It divides the

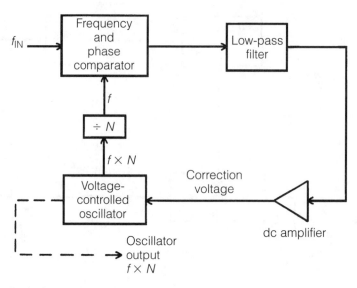

4-15 *A frequency synthesizer.*

VCO frequency by an amount equal to N. The value of N is set by a microprocessor.

Remember that the input to the comparator must be equal in phase and frequency to the incoming signal frequency (f). Therefore, the oscillator frequency must be $N \times f$ so that, when it is divided by N, the signal to the comparator will have a frequency equal to the incoming signal (f_{in}).

$$\frac{f \times N}{N} = f$$

You will find applications of the system in Fig. 4-14 in tuners of TVs and in the RF section of communications receivers.

Practice question

For the system shown in Fig. 4-15, the input frequency (f_{IN}) is 200 kHz. To get an oscillator frequency of 400 kHz, the $\div N$ counter must be set to:

A. 2
B. 4

Answer: Choice A is correct. If the oscillator operates at 400 kHz and its frequency is divided by 2 the output of the $\div N$ programmable counter will be 200 kHz. A frequency and phase lock occurs because that frequency equals f_{IN}.

Power supplies

Because an effort has been made to avoid using trade-name circuits that are not known to all technicians, you can expect circuits used in the CET test to be familiar to all types of technicians. The power supply is one circuit that everyone working in electronics should be familiar with, regardless of what system brand name they are working on. You might encounter a question about battery-type power supplies. You should understand that a primary cell cannot be recharged. A secondary cell can be recharged.

Primary cells can be rejuvenated. That is usually accomplished by heating them. Secondary cells are recharged by reversing the chemical process that occurs during discharge.

Secondary cells are recharged with a reverse current. That reverses the chemical process that occurs during discharge; that is, during the operation of the battery in supplying energy to a load resistance.

When batteries are connected in series, their voltages add. When they are connected in parallel, their currents add. A diode might be needed in each parallel branch as shown in Fig. 4-16. That prevents one battery from trying to recharge another.

Analog regulated power supplies

It is important to become familiar with the various rectifier configurations. You should be able to recognize half-wave rectifiers, full-wave rectifiers, bridge rectifiers, half-wave doublers, and full-wave doublers. These basic rectifier circuits are not covered here. If you encounter them in a Journeyman CET test, it will most likely be in conjunction with some other circuitry.

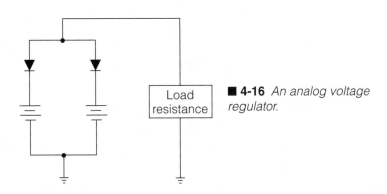

■ **4-16** *An analog voltage regulator.*

It is useful to understand the term *regulation* as it applies to an unregulated supply.

Regulation is a measure of how well the supply holds its output voltage under varying load conditions. The percent regulation of a power supply is given by the following equation:

$$\% \; regulation = \frac{No\text{-}load \; voltage - Full\text{-}load \; voltage}{Full\text{-}load \; voltage} \times 100$$

A low value of percent regulation is desirable. The term *percent regulation* is meaningless in a regulated supply. If the regulated power supply circuit is properly designed it will not be possible to get a measurable change in output voltage for changes in load current over the specified values given by the designer.

Figure 4-17 shows a simplified drawing of a closed-loop analog (continuous-voltage) regulator. This circuit can be divided into some basic functions.

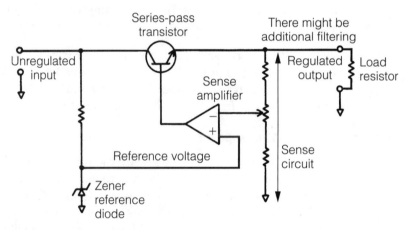

■ **4-17** *An example of a voltage regulator.*

The *sense circuit*, which has three resistors in series, is used to sense the output voltage of the power supply. In order for regulation to occur, the regulator circuit must "know" what the output voltage is. The sense circuit provides that information.

Another requirement in a regulated circuit is a reference voltage. It is usually obtained with a zener diode.

In the sense amplifier, the reference (or zener) voltage is compared with the sense voltage of the power supply. When those two voltages are the same, no correction voltage is delivered by the sense amplifier.

If the sense voltage is too high, the amplifier will deliver a bias voltage to the series-pass transistor, which lowers the current through the power-supply output resistance. That, in turn, will lower the power supply output voltage.

When the sense voltage is too low, the sense amplifier sends a bias voltage to the sense amplifier, which increases the power supply current. That will increase the power supply output voltage.

Observe that the resistors for the voltage-sense circuit in the circuit of Fig. 4-17 are connected in parallel with the power supply load resistance. If a current sense is needed, the sense circuit is connected in series with the power supply load resistance.

Figure 4-18 shows a current regulator with a series sense resistor. If the current through the resistor is too high, the voltage across this resistor is above 0.7 volt. That causes the silicon transistor to conduct. Its collector current flows through R and reduces conduction through the series-pass transistor.

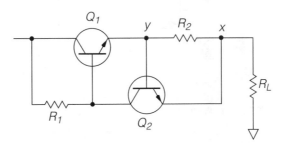

■ **4-18** *An increase in load current increases the voltage across R_2. That causes Q_2 to conduct harder through R, lowers the base current to Q_2, and reduces the load current.*

Remember that the term *load* in a power supply always refers to the current in the output. Do not confuse the *load* with the *load resistance*. Load resistance represents the opposition (the useful output resistance of a power supply), but the load is always the power-supply current.

Instead of using an analog feedback system it is possible to have a regulated power supply that uses a pulse for regulation. That type of supply is usually referred to as a *switching regulator*.

Figure 4-19 shows a block diagram of a switching regulator. It is another form of closed-loop power supply system. Switching regulators continually turn the load current ON and OFF. The amount of ON time is determined by the closed-loop regulating circuitry. If the output voltage or power is too low, then the load current is switched on for a longer period of time. Conversely, it is switched on for a shorter period of time if the output voltage (or load power) is too high. By regulating the ON time, it is possible to control the output voltage and output power of the supply.

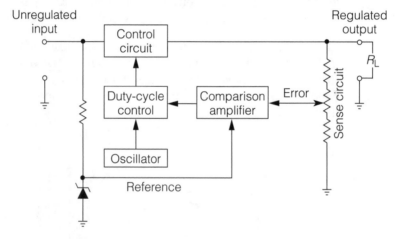

■ **4-19** *Block diagram of a switching regulator.*

Switching regulators are more efficient than the analog types. However, they cost more and they use more complicated circuitry.

Another power supply that you should be familiar with is the scan-derived type used in some television receivers. It obtains the ac voltage necessary for output dc power from the flyback transformer.

Because the output dc voltage from a scan-derived supply depends on a signal from the flyback transformer, and the signal from the flyback transformer comes from the amplifiers, which require a dc voltage, it is obvious that this system cannot start by itself. A special start-up circuit is necessary to start the horizontal oscillator in the flyback system. That, in turn, produces the necessary flyback energy which can be rectified and utilized in the scan-derived system.

The start-up circuitry is designed by the manufacturer. For that reason, a number of different circuits are being used. Spend some time reviewing those circuits before taking the Journeyman Consumer CET test.

Oscillators

Oscillators are divided into two groups: *sinusoidal* and *nonsinusoidal* (or relaxation). Both kinds are readily available in IC form. The timing capacitors and/or inductors for IC oscillators are likely to be located external to the IC package. Also, the oscillators in integrated circuits are often parts of a larger circuit.

Remember that an *oscillator* is a circuit that converts dc to ac. Circuits that convert ac to dc are *rectifiers*. *Converters* change dc from one value to another, and *inverters* change dc to ac. Although inverters and oscillators both change dc to ac, only inverters are used directly as power supplies.

Most inverters use an oscillator as part of their circuitry. However, that is not always the case. There are also inverters that are motor-generator combinations.

Sine-wave oscillators are often named after their inventors. Examples are: Armstrong, Hartley, and Colpitts. Relaxation oscillators are named for their circuitry. Examples are: free-running multivibrators, one-shot multivibrators and blocking oscillators. The multivibrators use RC timing networks and the blocking oscillators use transformer feedback as well as RC timing. These oscillator circuits are subjects in basic electronics. You might find it useful to review them before taking a test.

Protection circuits

Fuses and circuit breakers are the most common protection components. Always use exact replacement components for those devices. Before re-setting a circuit breaker or replacing a fuse, it is a good idea to check the circuit to find out why an overload occurred.

A *crowbar circuit* is shown in Fig. 4-20. It consists of an SCR connected across the load resistance in a power-supply system. It is a protective circuit and is used to prevent damage to ICs and other voltage-sensitive circuits from a temporary or permanent overvoltage. The rapid action of the SCR protects the output (R_L) from an overvoltage. Also, the high SCR current blows the fuse.

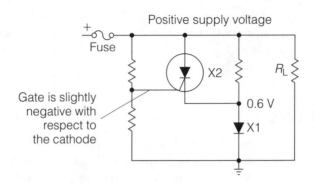

■ **4-20** *An example of a crowbar circuit.*

Practice question

In a certain television receiver, the ac input voltage to the rectifier comes from the flyback transformer. This is an example of:

A. a switching regulator.
B. a scan-derived supply.
C. a crowbar circuit.
D. an analog regulator.

Answer: Choice B is correct.

Tracking supplies

Tracking regulators are used when both positive and negative dc voltages are obtained from a single power supply. With the tracking regulator configuration, any increase in the positive voltage setting will be automatically accompanied by the same amount of change in the negative output voltage. As an example, if the positive output voltage is increased from 10 to 12 volts (by adjustment), the negative output voltage will change from –10 to –12 volts automatically.

Practice questions

1. Feedback in a blocking oscillator is through a:
   A. coupling capacitor.
   B. diode.
   C. transformer.
   D. saturated transistor.
   E. coaxial cable.

2. When the base of a transistor is the same voltage as the emitter, the transistor is:

   A. saturated.
   B. operating normally.
   C. cut off.

3. The free-running frequency of a synchronized oscillator should be:

   A. exactly equal to the synchronizing frequency.
   B. slightly below the synchronizing frequency.
   C. slightly above the synchronizing frequency.

4. The greatest portion of a superheterodyne receiver's gain is accomplished in the:

   A. RF amplifier.
   B. IF amplifier.
   C. detector stage.
   D. video amplifier.

5. Which of the following FM detector circuits is insensitive to amplitude modulation?

   A. Foster-Seeley FM discriminator
   B. Envelope detector
   C. Ratio detector
   D. Slope detector
   E. Homodyne

6. A circuit in which a single amplifier acts as both a sound IF amplifier and an audio amplifier is called:

   A. an impossibility.
   B. a reflex-amplifier circuit.
   C. a reflectodyne circuit.
   D. a neutrode circuit.

7. Radiation of the local-oscillator signal from a superheterodyne receiver is prevented or greatly reduced by the use of:

   A. reducing the dc voltage to a local oscillator.
   B. twisting the twin-lead line from the antenna.
   C. an RF amplifier.
   D. forward AGC.

8. An advantage of using an FET over a bipolar transistor is that:

   A. no input signal power is needed.
   B. it can be used in a Darlington configuration.
   C. it can be used in a complementary amplifier configuration.
   D. it produces no crossover distortion in push-pull amplifiers.

9. The IF frequency of an FM broadcast receiver is:
   A. 41.25 MHz.
   B. 10.7 MHz.
   C. 3.58 kHz.
   D. 455 kHz.
   E. 4.5 kHz.

10. Which of the following is the IF frequency for a standard AM broadcast receiver?
   A. 10.7 MHz
   B. 10.7 kHz
   C. 540 kHz
   D. 0.455 MHz

11. When the television primary colors are combined in the color tube so that $V_y = 0.59\,V_G + 0.30\,V_R + 0.11\,R_B$ the result is:
   A. a white screen.
   B. a green color on the screen.
   C. a black screen.

12. Another name for a Darlington connection is:
   A. alpha squared.
   B. beta squared.
   C. gamma squared.
   D. delta squared.

13. An integrator is a:
   A. circuit that combined two different kinds of oscillator operations.
   B. low-pass filter.
   C. circuit that combines two different kinds of pulses.
   D. type of delay line.

14. A ferrite bead behaves like:
   A. an inductor.
   B. a temperature-sensing resistor.

15. The filter circuit shown in Fig. 4-21 is a:
   A. high-pass type.
   B. low-pass type.

16. For the amplifier shown in Fig. 4-22, the output signal is:
   A. in phase with the input signal.
   B. 180° out of phase with the input signal.

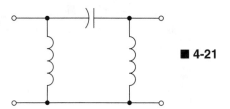

■ 4-21

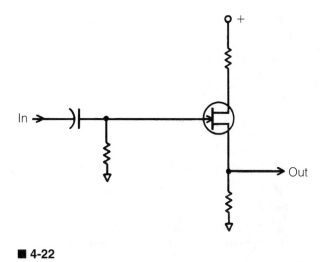

In →

+

Out

■ 4-22

17. Which amplifier configuration will introduce a 180° phase shift to the signal?

    A. Common drain
    B. Cathode follower
    C. Grounded grid
    D. Common source

18. Which amplifier configuration can be used to match a high impedance to a low impedance?

    A. Grounded base
    B. Conventional
    C. Emitter follower
    D. Common source

19. You would expect to obtain the best high-frequency response from a:

    A. conventional amplifier.
    B. common-base amplifier.
    C. common-drain amplifier.
    D. grounded-source amplifier.

20. Disregarding the forward emitter voltage drop, class-B operation of a bipolar transistor amplifier is best obtained by using:

 A. base-leak bias.
 B. no forward bias.
 C. emitter resistor bias.
 D. voltage divider bias.

21. Which amplifying device would normally produce the greatest amount of noise in an RF amplifier circuit?

 A. MOSFET amplifier
 B. Bipolar transistor amplifier

22. Which type of bias could result in a destroyed bipolar transistor if the input signal is lost?

 A. Voltage divider bias
 B. Power supply bias
 C. Class-B bias
 D. None of these choices is correct.

23. Removing the capacitor across an emitter resistor in a transistor amplifier will:

 A. increase the gain of the amplifier.
 B. increase the amplifier frequency response.

24. The emitter resistor in a bipolar transistor circuit is used to:

 A. reduce amplifier noise.
 B. increase the gain of the stage.
 C. bias the transistor.
 D. stabilize the stage against the effects of temperature change and against thermal runaway.

25. A bootstrap connection for an amplifier input circuit is used to:

 A. decrease the frequency response of the amplifier.
 B. increase the input impedance of the amplifier.
 C. decrease the output impedance of the amplifier.
 D. increase the output impedance of the amplifier.

26. Peaking compensation is used to:

 A. decrease bandwidth.
 B. improve fidelity of audio amplifiers.
 C. reduce noise.
 D. increase the high-frequency gain of an amplifier.

27. A common-collector amplifier has a gain of:

 A. less than 1.0.
 B. exactly 1.0.
 C. greater than 1.0.

28. The circuit in Fig. 4-23 is:
    A. an AM detector.
    B. a product detector.
    C. a half-wave voltage doubler.
    D. a full-wave voltage doubler.

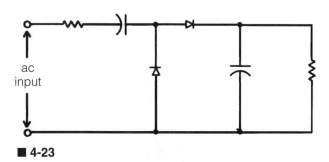

■ **4-23**

29. Why is a half-wave voltage doubler sometimes preferred over a full-wave doubler?
    A. It has higher output voltage.
    B. It has higher output current.
    C. It has no zero-volt line.
    D. It can be grounded to earth.

30. With a sine-wave voltage input to a differentiating (RC) circuit, the output voltage waveform should be a:
    A. sine waveform.
    B. sawtooth waveform.
    C. dc voltage.
    D. square wave.

31. With a sine-wave voltage input to an integrator circuit, the output voltage waveform should be a:
    A. sine wave.
    B. sawtooth wave.
    C. dc voltage.
    D. square wave.

32. In a parallel LC circuit, the resonant frequency can be increased by:
    A. moving the capacitor plates closer together.
    B. moving the capacitor plates further apart.

33. Without changing the input resistor you can increase the gain of an operational amplifier circuit by:

    A. increasing the resistance of the feedback resistor.

    B. decreasing the resistance of the feedback resistor.

34. A possible problem with a push-pull power amplifier is crossover distortion. That problem:

    A. does not exist with a complementary power amplifier.

    B. also exists with a complementary power amplifier.

35. The strong mechanical connection shown in Fig. 4-24:

    A. should not be used prior to soldering.

    B. should be used prior to soldering.

■ 4-24

36. Which of the circuits in Fig. 4-25 is called a *differentiator*?

    A. The one marked A.

    B. The one marked B.

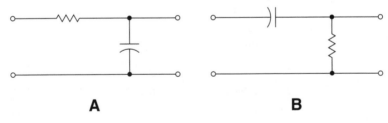

A　　　　　　　　　　**B**

■ 4-25

37. Which of the following diodes is normally operated with reverse bias?

    A. Zener

    B. Varactor

    C. Both A and B

    D. Neither

38. Increasing the bandwidth of an RF amplifier will:

    A. decrease its output noise.
    B. not affect its output noise.
    C. increase its output noise.

39. Consider a television receiver with no antenna connected to its antenna input terminals. Connecting a high-resistance resistor across its antenna terminals will:

    A. increase the noise output of the RF amplifier.
    B. not affect the noise output of the RF amplifier.
    C. decrease the noise output of the RF amplifier.

40. Which of the following cannot be used to demodulate an AM signal?

    A. Class-A amplifier
    B. Class-B amplifier

41. The voltage gain of the op-amp circuit in Fig. 4-26 is:

    A. $\dfrac{R_f}{R_i}$

    B. $\dfrac{R_i}{R_f}$

    C. $1 + \left(\dfrac{R_f}{R_i}\right)$

    D. $1 + \left(\dfrac{R_i}{R_f}\right)$

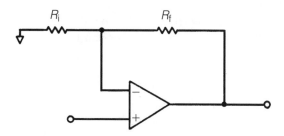

■ **4-26** *How is a bass tone emphasized?*

42. In the simple tone control of Fig. 4-27, an increase in bass tone is achieved by moving the arm of R:

    A. toward X.
    B. toward Y.

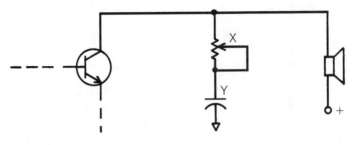

■ 4-27

43. The rise time of a square wave is the time it takes to go from:

   A. minimum V to maximum V.
   B. 10% of maximum V to 90% of maximum V.
   C. Neither choice is correct.

44. One problem with direct-coupled amplifiers is:

   A. poor frequency response.
   B. low gain.
   C. excessive noise.
   D. level shifting.

45. Which of the following amplifier configurations has a low input impedance and a high output impedance?

   A. Common emitter
   B. Common base
   C. Common collector

46. In the partial amplifier circuit (shown in Fig. 4-28) closing the switch will:

   A. increase the amplifier bandwidth.
   B. decrease the amplifier bandwidth.

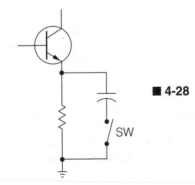

■ 4-28

Answers to the practice questions

Question	Answer
1.	C
2.	C
3.	B
4.	B
5.	C
6.	B
7.	C
8.	A
9.	B
10.	D
11.	A
12.	B
13.	B
14.	A
15.	A
16.	A
17.	D
18.	C
19.	B
20.	B
21.	B
22.	C
23.	B
24.	D
25.	B
26.	D
27.	A
28.	C
29.	D
30.	A
31.	A
32.	B
33.	A
34.	B
35.	A
36.	B
37.	C
38.	C
39.	A
40.	A
41.	C

Question	Answer
42.	A
43.	B
44.	D
45.	B
46.	B

Digital

Associate level *There is no digital section in the Associate-level CET test. However, questions on digital theory and applications can be used in any of the eight sections.*

Journeyman level *This material should be reviewed before taking the Journeyman Consumer Electronics, Industrial Electronics, or Computer Option tests.*

Numbers and counting systems

The radix of a numbering system is also known as the *base* of that system. It tells the number of symbols used to make all numbers in that system. For example, the radix of the decimal system is 10 because there are 10 symbols (0 through 9) used to make all of the numbers in that system.

Table 5-1 lists counts for numbers of five different radix values. It is especially important to be able to convert between the numbers in that table. The radix values are: binary (base 2), octal (base 8), decimal (base 10), hexadecimal (base 16), and BCD (Binary Coded Decimal (BCD), which is a special application of base 2).

Table 5-2 gives an example of conversion from binary to decimal and from decimal to binary numbers. If you do not remember how to do that for all two-system combinations in Table 5-1, it would be a good idea to spend some review time in a basic digital book.

Practice question

Complete the following bottom row for Table 5-1.

Answer: Reading from left to right the numbers are:

17	10001	21	11	0001 0111

Observe that in each case a new column is started after all of the symbols have been used. For example, all of the symbols have been used when the number 9 is reached in the decimal column (columns are vertical and rows are horizontal). So, the next number after 9 must be 10 with the number 1 starting a new column.

■ Table 5-1 Counting for different radix values.

Decimal	Binary	Octal	Hexidecimal	BCD
0	0000	0	0	0000
1	0001	1	1	0001
2	0010	2	2	0010
3	0011	3	3	0011
4	0100	4	4	0100
5	0101	5	5	0101
6	0110	6	6	0110
7	0111	7	7	0111
8	1000	10	8	1000
9	1001	11	9	1001
10	1010	12	A	0001 0000
11	1011	13	B	0001 0001
12	1100	14	C	0001 0010
13	1101	15	D	0001 0011
14	1110	16	E	0001 0100
15	1111	17	F	0001 0101
16	10000	20	10	0001 0110
17	- - - -	- - - -	- - - -	- - - - - - - -

Practice question

Convert decimal number 235 to a BCD number.

Answer: Write each digit of the decimal number as a binary number.

2	3	5
0010	0011	0101

The BCD equivalent to 235 is 001000110101. That is better expressed as 0010 0011 0101.

As another example of starting a new column in a count, refer again to the binary column in Table 5-1. All of the symbols (0 and 1) have been used in the binary column when the number 0001 is reached. Therefore, a new column must be started to get 0010.

Practice question

Without looking at the table, what is the next binary number after 1011?

Answer: Reading from right to left all of the symbols have been used in the first two columns. So, a binary 1 must be used to start a third column: 0100. Disregard the zeros in front of the numbers in the binary and BCD columns. They are only used as place setters.

It is customary to write the radix of a number as a subscript in cases where there might be some confusion. For example, 10 could be a

■ Table 5-2 Conversion examples.

Decimal-to-binary conversion

This conversion requires that the decimal number be repeatedly divided by 2 and the remainder written for each step. Example: convert decimal 13 to a binary number

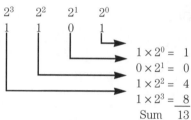

Division   Remainder

$13 \div 2 = 6$ 1

$6 \div 2 = 3$ 0

$3 \div 2 = 1$ 1

$1 \div 2 = 0$ 1

Read the binary equivalent

Decimal 13 = 1101

Binary-to-decimal conversion

A binary number is written with each digit in a place factor. Here are the first six place factors

$$\leftarrow -\bullet- \quad 2^6 \; 2^5 \; 2^4 \; 2^3 \; 2^2 \; 2^1 \; 2^0$$

To make a binary-to-decimal conversion, write the binary number. Then, multiply the binary number by the place factor. Add the products. Example: convert binary 1101 to a decimal number

Place factors 2^3 2^2 2^1 2^0

Binary number   1   1   0   1

$1 \times 2^0 = \quad 1$

$0 \times 2^1 = \quad 0$

$1 \times 2^2 = \quad 4$

$1 \times 2^3 = \underline{\quad 8}$

Sum   13

Answer: binary 1101 = Decimal 13

decimal, binary or hexadecimal number. Here are the ways numbers in each row can be written using subscripts (refer to row 13):

12_{10} 1100_2 14_8 C_{16} BCD: 001 0010

The five methods of counting shown in Table 5-1 are very important for understanding the counters used in digital systems. Again, you should know all of them and be able to convert between them.

Practice question

Replace the Xs with a number in the following problem:

$$14_{10} = XXXX_{16}$$

A. F

B. 14_{16}

C. E_{16}

Answer: Choice C is correct (see Table 5-1).

The subscript identifies the base of the numbering system. So, E_{16} means that the number is hexadecimal. The answer to the question can be obtained from Table 5-1. However, you are likely to be asked questions about conversions of higher numbers. Conversion problems in the CET test involve numbers that are less than 32_{10}.

Practice question

Complete the row starting with decimal 18 below:

Answer: The correct numbers for the row are:

18	___	___	___	___	___

Answer: 10010   22   12   0001 1000

Boolean algebra

There might be a few questions in a CET test that require a basic knowledge of Boolean algebra. The laws of Boolean algebra are given in Table 5-3. Boolean algebra is useful for determining the output of a combinational logic circuit. An example is given in Table 5-4.

■ **Table 5-3 Laws of Boolean algebra.**

1. $A + 1 = 1$	OR
2. $A + 0 = A$	OR
3. $A + A = A$	OR
4. $A(1) = A$	AND
5. $A(0) = 0$	AND
6. $A \times A = A$	AND
7. $A \times \overline{A} = 0$	AND
8. $A + \overline{A} = 1$	OR
9. $A + \overline{A}B = A + B$	OR
10. $(A + B)(A + C) = A + BC$	AND/OR
11. $\overline{A + B} = \overline{A} \times \overline{B}$	(DeMorgan's law)
12. $\overline{A \times B} = \overline{A} + \overline{B}$	(DeMorgan's law)
13. $\overline{A}B + A\overline{B} = A \oplus B$	Exclusive OR
14. $(\overline{A} \times \overline{B}) + (A \times B) = \overline{A \oplus B}$	Exclusive NOR
	also called logic comparator

By writing the Boolean expression for the output of each circuit, it is sometimes possible to reduce the number of components in that circuit. See Table 5-4.

■ **Table 5-4 Sample Problem.**

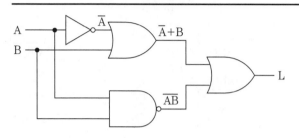

$$L = \overline{A} + B + \overline{AB}$$

By DeMorgan's theorem: $\overline{AB} = \overline{A} + \overline{B}$

$$L = \overline{A} + B + \overline{A} + \overline{B}$$

Combining like terms:

$$L = \overline{A} + \overline{A} + B + \overline{B}$$

By Boolean laws

$$\overline{A} + \overline{A} = \overline{A}$$
$$B + \overline{B} = 0$$

So, the output is: $L = \overline{A} + 1 = 1$

Replace with:

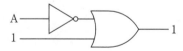

Note: You cannot replace the circuit
with logic 1 because it may ruin
system propagation delay.

Symbols and other identifications of gates

You should not take any CET test unless you have a very good understanding of the basic logic gates. Specifically, you should know the truth table for each gate, the symbols used for the gates, the Boolean equation for each gate, and you should be able to recognize a simple circuit that represents the basic gates (Table 5-5).

117

■ Table 5-5   Truth tables, symbols, Boolean equations, and a representative circuit.

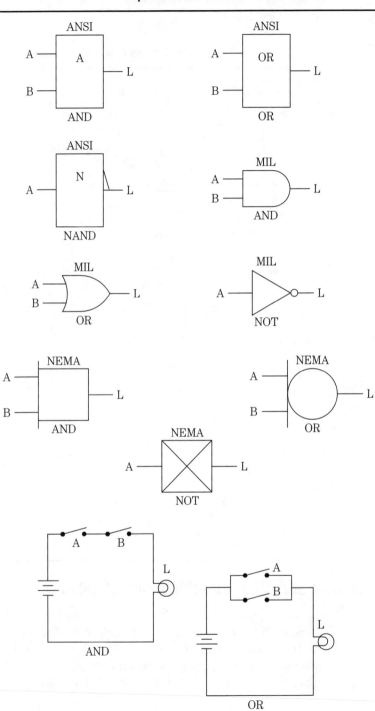

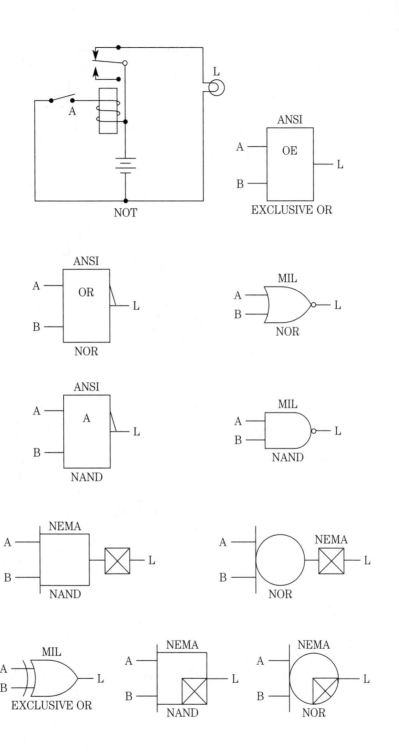

NOT

ANSI
A — OE — L
B —
EXCLUSIVE OR

ANSI
A — OR — L
B —
NOR

MIL
A —
B —
NOR — L

ANSI
A — A — L
B —
NAND

MIL
A —
B —
NAND — L

NEMA
A —
B —
NAND — L

NEMA
A —
B —
NOR — L

MIL
A —
B —
EXCLUSIVE OR — L

NEMA
A —
B —
NAND — L

NEMA
A —
B —
NOR — L

119

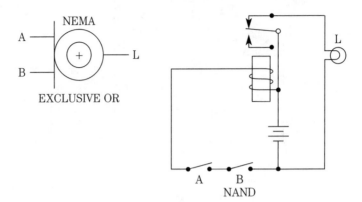

NEMA

EXCLUSIVE OR

NAND

$A \times B = L$
AND

A	B	L
0	0	0
0	1	0
1	0	0
1	1	1

OR $A + B = L$
Also called
INCLUSIVE OR

A	B	L
0	0	0
0	1	1
1	0	1
1	1	1

$A = \overline{L}$
NOT

A	L
0	1
1	0

$\overline{A \times B} = L$
NAND

A	B	L
0	0	1
0	1	1
1	0	1
1	1	0

$\overline{A + B} = L$
NOR

A	B	L
0	0	1
0	1	0
1	0	0
1	1	0

$\overline{A}B + A\overline{B} = L$
also
$A \oplus = L$
EXCLUSIVE OR

A	B	L
0	0	0
0	1	1
1	0	1
1	1	0

$\overline{AB} + AB = L$
EXCLUSIVE NOR
(logic comparator)

A	B	L
0	0	1
0	1	0
1	0	0
1	1	1

You must understand those things for each of the following basic logic gates: AND, OR, NOT, NAND, NOR, inverter, exclusive OR, exclusive NOR, tri-state buffers, and tri-state inverters.

Inverters are also called a *NOT gates*. The term *half-adder* is sometimes used in place of *exclusive OR*. An *exclusive NOR* is also called a *logic comparator*.

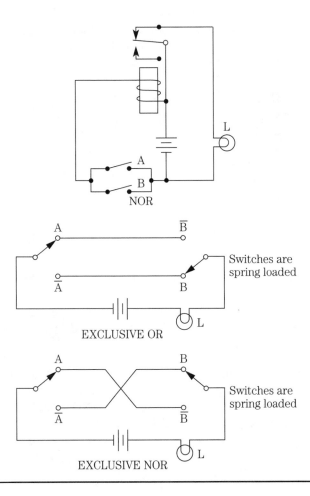

NOR

EXCLUSIVE OR

Switches are
spring loaded

EXCLUSIVE NOR

Switches are
spring loaded

The four essential characteristics of the basic logic gates are given in Table 5-5. Make sure that you can identify all of the characteristics given in that table!

The truth tables given in Table 5-5 are for two-input logic gates. From those tables, you should be able to identify truth tables for three-inputs, four-inputs, or from any number of inputs.

For example, look at the truth table for the AND gate. Note that the only way to get an output level of LOGIC 1 is to have all of the inputs at a LOGIC 1 level. So, in a truth table for a three-input AND gate all three inputs would have to be a LOGIC 1 level in order to get a LOGIC 1 level at the output (L), output.

Notice also that the only way you can get a logic 0 output from the OR gate is to have all the inputs at logic 0. So, if it is a four-input

OR gate, all four inputs would have to be at LOGIC 0 in order to get a LOGIC 0 level at the output. This same reasoning can be applied to the other gates.

Practice question

Identify the following truth table:

A	B	C	L
0	0	0	1
0	0	1	1
0	1	0	1
0	1	1	1
1	0	0	1
1	0	1	1
1	1	0	1
1	1	1	0

Answer: It is the truth table for a three-input NAND gate. The only way to get a LOGIC 0 out of a NAND gate is to have all inputs at a LOGIC 1.

Table 5-5 also gives the Boolean equations for each of the basic gates. In some cases, different equations can be used to represent the same gates. In the case of the AND gate, "A AND $B = L$" can be expressed in three different ways:

$$A \times B$$
$$A \bullet B$$
$$AB$$

You might encounter a Boolean Logic equation in any of the forms given in the table.

Practice question

Identify the gate that has the following Boolean expression:

$$\overline{A}B + A\overline{B} = L$$

Answer: The Boolean equation is for an EXCLUSIVE OR. Read it as follows: NOT A AND B OR A AND NOT B equals L.

Here is a summary of some important characteristics of the basic gates:

☐ For the NAND gate, several equations can be used. Observe that an overbar can be placed on any of the three expressions above for AND to obtain a NAND.

- The exclusive OR gate is represented two different ways with Boolean equations.
- Three different symbols are used for the gates in Table 5-5. They are identified as: conventional, NEMA, and ANSI.
- The symbols used in the Associate-level and Journeyman-level CET tests use conventional symbols. The other symbols are given for your reference.

There doesn't seem to be any shortcut method of memorizing the basic Boolean expressions. If you have worked with digital electronics it will not be difficult to recognize any one of them in a CET test. However, if this subject is fairly new to you, memorize all of the versions given in the table.

Important rules for overbars

You should also know a few rules for overbars as they relate to Boolean algebra. Any time there is an even number of overbars, they can be eliminated without changing the meaning of the Boolean expression.

If there is an odd number of overbars, you can eliminate pairs of overbars without changing the value of the Boolean expression. Examples for these rules are shown in Table 5-6. It is useful to know the rules for overbars when you are simplifying Boolean expressions.

■ **Table 5-6 Important examples of the rules for overbars.**

Any time there is an even number of overbars the expression can be written without any overbars.

$$\bar{\bar{A}} = A$$

$$\bar{\bar{\bar{A}}} = A$$

Any time there is an odd number of overbars the expression can be written with one overbar.

$$\bar{\bar{\bar{A}}} = \bar{A}$$

$$\bar{\bar{\bar{\bar{A}}}} = \bar{A}$$

Common errors using DeMorgan's theorem.

$\overline{AB} = \bar{A} \cdot \bar{B}$	Wrong!
$\overline{AB} = \bar{A} + \bar{B}$	Correct
$\overline{A + B} = \bar{A} + \bar{B}$	Wrong!
$\overline{A + B} = \bar{A} \cdot \bar{B}$	Correct

DeMorgan's theorems

You should know the two Boolean equivalents that are referred to as DeMorgan's theorems. They are listed in Table 5-3. The terms on the left side of the equation are identical to those on the right side of the equation, so those equalities might be referred to as identities.

Technicians use the following mnemonic device "break the bar, change the sign" for converting an expression on the left side of the equation to an expression on the right side of the equation. DeMorgan's rules are very useful for simplifying Boolean equations. They are also valuable for simplifying basic circuits. Table 5-7 shows common errors related to DeMorgan's theorems. Do not make those errors!

■ Table 5-7
Errors in
DeMorgan's theorem

Wrong:	$\overline{AB} = \overline{A} + \overline{B}$
Right:	$\overline{AB} = \overline{A} + \overline{B}$
Wrong:	$\overline{A + B} = \overline{AB}$
Right:	$\overline{A + B} = \overline{A}\,\overline{B}$

To prove the equations are not true, try to apply the rule of break the bar, change the sign to either side of a wrong equation in Table 5-6 and you will see why they are not true!

Some basics on hardware

It is presumed that you understand how to count the pins on an IC, regardless of whether it is a flat pack, a DIP package, or a TO-5 case. Always remember that the pins are counted counterclockwise when the devices are viewed from above and clockwise when it is viewed from below. This same rule applies to counting pins in tube sockets, relay sockets, and other electronic components.

To count the pins, always look for an identifying mark, such as a dot or tab (or a space between the pins). That tells you where to start the count. Again, it is important to remember that when you are looking at the top of the component, you must count counterclockwise.

Practice question

The arrow in Fig. 5-1 is pointing to pin number:

A. 1.
B. 14.

Answer: Choice B is correct. You are looking at the top of the IC package, so the count is counterclockwise. Pin 1 is nearest to the observer starting at the notch.

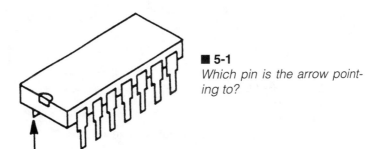

■ 5-1
Which pin is the arrow pointing to?

Practice question

The arrow in Fig. 5-2 is pointing to pin number:

A. 4.
B. 11.

Answer: A

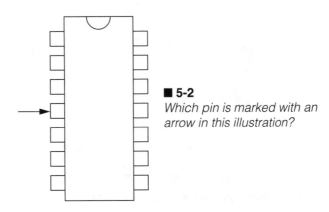

■ 5-2
Which pin is marked with an arrow in this illustration?

Use of three-state devices

Figure 5-3 shows two important *three-state* (or *tri-state*) *devices* that are essential to your knowledge of basic gates. The devices are three-state buffers and three-state inverters. The purpose of those devices is to provide isolation between the input and output

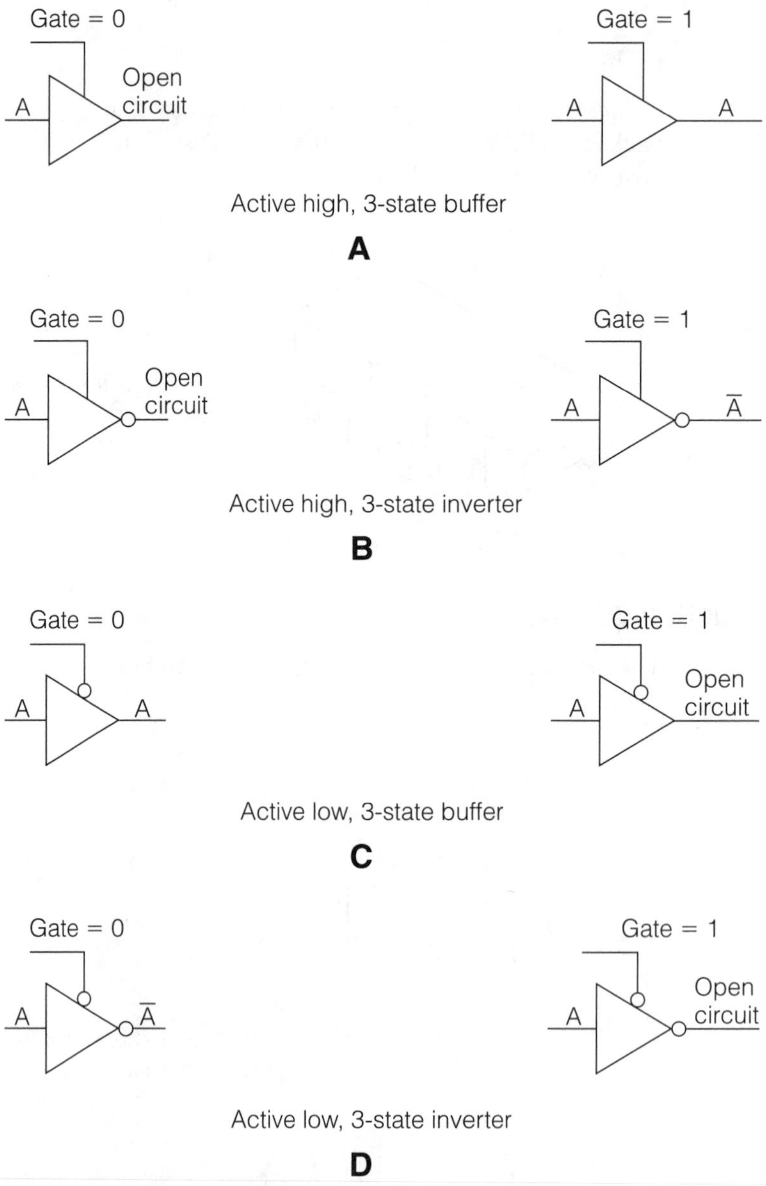

Active high, 3-state buffer

A

Active high, 3-state inverter

B

Active low, 3-state buffer

C

Active low, 3-state inverter

D

■ **5-3** *Two important three-state devices.*

terminals when the gate lead logic level is properly applied. Figure 5-4 illustrates a use of a tri-state device.

When the three-state terminal in Fig. 5-3A goes to a logic 1, the input and output (A and L) are coupled as in a normal buffer. There is an important exception. Any time you see a small circle where a lead enters a logic symbol, it means that the input is active low. In other words, a logic level 0 is required to activate the device.

The example in Fig. 5-3A is called a *three-state buffer*. It does not change the signal between the input and output. In some cases, especially in linear circuits, buffers might be used to increase the voltage or current as necessary for a particular design. In Figs. 5-3C and 5-3D, the same devices are shown with logic low at the input gate.

Buffers can also be used to introduce a very short time delay. That delay is equal to the propagation delay of the device. Propagation delay is defined as the time it takes for a signal to pass through a device.

The logic level of the input to a buffer is not changed. So, an input level of logic 1 will result in an output of logic 1. An input of logic 0 will give an output of logic 0.

127

The gate shown in Fig. 5-3B is a three-state inverter. As with the three-state buffer, there is no output from this gate when the three-state terminal is at logic 0. When the three-state terminal is changed to a logic 1, whatever level is on the input terminal will be

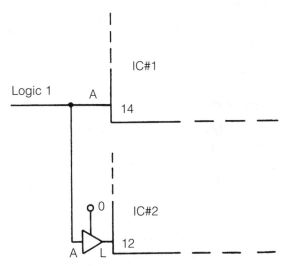

■ **5-4** *Examples of uses for tri-state devices.*

inverted at the output. Three-state devices are very important on bus lines in microprocessor systems.

Practice question

Supply the outputs for the two-way bus assuming the conditions shown in Fig. 5-5.

Answer: A - 1
        B - OPEN
        C - 1
        D - 1

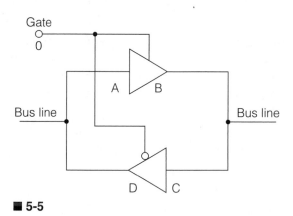

■ 5-5

Practice question

What is the logic level on pin 12 of IC number 2 in Fig. 5-4 as shown?

A. Logic 1
B. Logic 0
C. Neither choice is correct.

Answer: Choice C is correct. Actually, the output of the three-state buffer is an open circuit. That is not the same as saying logic 0! In practice, the terminal might float to a logic 1 or 0, but from the information in the drawing, you could not say that choice A or choice B is correct.

It is possible in some systems for two signals to appear on a bus line at the same time. That can cause a damaging short circuit if one of the signals is at a logic 1 level and the other signal is at a logic 0 level.

By using three-state devices the output of one gate will not be affected by the output of another gate on the same bus line. A bus is a combination of conducting wires or printed circuit conductors that transfer logic signals from one point to another.

An example of a three-input gate application is shown in Fig. 5-5. The circuit is for a two-way bus. It allows a logic signal to go from left to right or from right to left. You can see that a signal cannot flow in both directions at the same time. The allowed direction depends upon whether the three-state input is at logic level 1 or logic level 0.

Practice question

For the two-way bus in Fig. 5-6 the gate input is at logic level 0. This means that a signal on the bus can move from:

A. right to left.
B. left to right.

Answer: Choice A is correct. The buffer conducts because its gate input is active low. The other gate has an active high input. Therefore, a logic level 0 will not activate it.

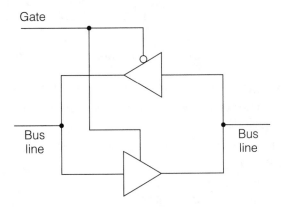

Gate

Bus line

Bus line

■ **5-6** *Which way will the signal go?*

Flip-flops

There are two important terms used in logic systems that you should remember. Static logic systems do not use clock signals. A clock signal IS employed in dynamic logic systems—also called *active systems*. Clock signals are pulses or rectangular waves that are used for timing and operating logic systems and microprocessors.

Flip-flops are components that are used in both static and dynamic systems. They are valuable in digital and microprocessor systems because they have two stable states: high and low. Other terms used for the two stable states are *SET* and *RESET*, and *ON* and *OFF*.

Figure 5-7 illustrates the two stable states for a flip-flop. This is an illustration for all types of flip-flops. As shown in Fig. 5-7, a flip-flop is in a HIGH (or SET or ON) condition when its "Q" output terminal is at logic 1 and its NOT Q output terminal ($\overline{Q}$) is at logic 0.

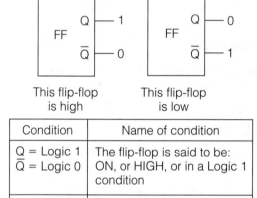

Condition	Name of condition
Q = Logic 1 $\overline{Q}$ = Logic 0	The flip-flop is said to be: ON, or HIGH, or in a Logic 1 condition
Q = Logic 0 $\overline{Q}$ = Logic 1	The flip-flop is said to be: OFF, or LOW, or in a Logic 0 condition

■ **5-7** *Two stable states for a flip-flop.*

A flip-flop is in a LOW (or, RESET or OFF) condition when its "Q" output terminal is at logic 0 and its NOT Q output terminal ($\overline{Q}$) is at logic 1.

There are many other components that have two stable states. A few examples are:

☐ switches are closed or open.
☐ lamps are ON or OFF.
☐ relays are energized or de-energized.
☐ transistors are conducting or cut off.

The popularity of two-state devices can be explained on the basis of the binary counting system. There are only two digits (0 and 1) in the binary system and two-state devices can easily be used to represent the binary numbers. By contrast, there are fewer devices with 10 stable states that could be used to represent the decimal numbers.

RS flip-flops

An important type of flip-flop is illustrated in Fig. 5-8. It is called an *RS (Reset and Set) flip-flop*. As shown in the illustration, this type of flip-flop can be made with either NAND or NOR gates. The way it is constructed affects its operation in a circuit. It is best understood by carefully studying the various flip-flop conditions illustrated in Fig. 5-9. They are the conditions for a NAND type. A similar set of drawings can be made for a NOR flip-flop.

You should know that there are theory purists who say the term *flip-flop* should not be used to describe the NAND and NOR devices discussed in this section. They feel it should properly be referred to as a *latch* or *data latch*. You will see those terms in technical literature. However, you will also see these devices referred to as *flip-flops*.

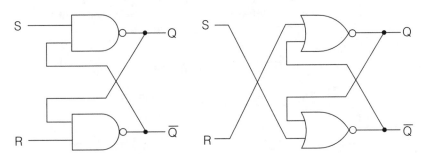

R-S flip-flop made with NANDs   R-S flip-flop made with NORs

NAND latch
truth table

S	R	Q	Q̄	
1	1	1	0	High
1	0	0	1	Change to low
1	1	0	1	Low
0	1	1	0	Change to high
1	1	1	0	High
0	1	1	0	High
1	1	1	0	High

■ **5-8** *RS flip-flop.*

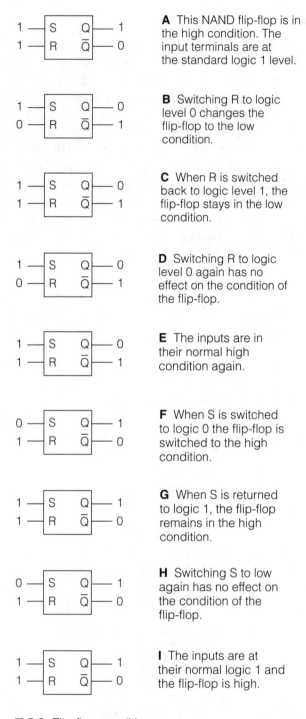

A This NAND flip-flop is in the high condition. The input terminals are at the standard logic 1 level.

B Switching R to logic level 0 changes the flip-flop to the low condition.

C When R is switched back to logic level 1, the flip-flop stays in the low condition.

D Switching R to logic level 0 again has no effect on the condition of the flip-flop.

E The inputs are in their normal high condition again.

F When S is switched to logic 0 the flip-flop is switched to the high condition.

G When S is returned to logic 1, the flip-flop remains in the high condition.

H Switching S to low again has no effect on the condition of the flip-flop.

I The inputs are at their normal logic 1 and the flip-flop is high.

■ **5-9** *Flip-flop condition.*

132

Here are some points about the operations shown in Fig. 5-9 that you should keep in mind.

☐ When the NAND flip-flop is in a HIGH condition it is stable only if logic 1 levels are delivered to the R and S input terminals. In order for a NOR flip-flop to be stable, logic 0 levels must be delivered to the input terminals.

☐ The only way you can get an RS flip-flop to change condition is to deliver opposite logic levels to the input (RS) terminals.

☐ The illustration shows that a logic 0 is delivered to one of the terminals to change its condition. For a NOR flip-flop, it is necessary to deliver a logic 1 to one of its terminals to change its condition.

☐ Observe carefully that the condition of a NAND flip-flop can only be changed to HIGH when the input logic 0 is on the proper terminal. For example, you cannot change a flip-flop to a high condition by delivering logic 0 to S if it is already in the high condition.

Practice question

A certain RS flip-flop made with NOR gates has a logic level 1 delivered to both input terminals. That flip-flop is:

A. in a stable condition.
B. in an unstable condition.

Answer: Choice B is correct. When a flip-flop is in an unstable condition, its output cannot be predicted. In other words, it can be either HIGH or LOW.

An RS flip-flop can be used to store bits in memory. The flip-flop is placed in a 1 or 0 condition (HIGH or LOW), depending on which digit (1 or 0) is to be memorized. It will hold that position for any desired length of time. Because these flip-flops operate with a power supply, their stable condition is only true as long the power supply voltage is not interrupted. However, if the power supply voltage is removed, the stored number is lost.

This type of memory component is referred to as being *volatile*. A nonvolatile memory will retain the stored information—even though the power supply is turned off.

Toggled flip-flops

Figure 5-10 shows the symbol for a toggled flip-flop. It changes its output every time the leading edge of a clock input arrives. As with

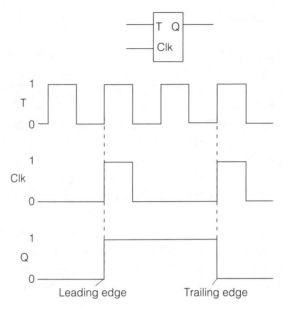

The clock positive transition
and toggle positive transition
produce a Q transition

■ **5-10** *Signals for a toggled flip flop.*

all flip-flops, the output of the toggled flip-flop is HIGH if Q is at a logic 1 level and it is LOW when Q is at a logic 0 level.

The flip-flop in Fig. 5-10 divides the input frequency by two because it can only change its output condition during the leading edge (0 to 1) of the clock signal. There are also toggled flip-flops that change their output only during the trailing edge (1 to 0) of the clock signal.

The input (T) and output (Q) of the toggled flip-flop is compared with the clock signal in Fig. 5-10. This type of display is called a *timing diagram*. You can see from the diagram that a transition in the output only occurs when the clock signal goes from 0 to 1. Because of that characteristic the output signal at Q is at half the frequency of the clock signal.

Because the output frequency is half the input frequency, this arrangement is called a *divide-by-two circuit*. Many of the available types of flip-flops can be wired in such a way that they toggle with an input clock signal.

D flip-flops

Toggled flip-flops (Fig. 5-11) are very important in both passive and active logic systems. They are used in many counting systems.

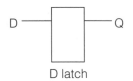

D latch

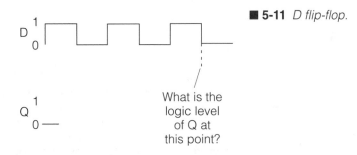

■ **5-11** *D flip-flop.*

One important application uses a D flip-flop as a memory compo- nent or memory cell. The *D* stands for data and this flip-flop can be used to store data in binary form. The characteristic of a D flip-flop that makes it useful in that application is its ability to remain in whatever condition it is placed until an imbalance in its input changes its output condition.

Practice question

Refer to the data flip-flop in Fig. 5-11. It is triggered when the in- put clock signal goes from high to low. The data stored in the D flip-flop is:

A. logic level 1.
B. logic level 0.

Answer: Choice A is correct. The high-to-low transition switches the flip-flop from low to high. The following low-to-high transition on the data input terminal will not change the condition of the flip-flop.

J-K flip-flops

The letters J and K that identify this type have no particular mean- ing. One disadvantage of the RS flip-flop is that it has a possible "not allowed" condition. Another disadvantage is that it is difficult (but, not impossible) to set it up for toggling. The J-K flip-flop does not have either of those disadvantages. Figure 5-12 shows the symbol and truth table of a J-K flip-flop. Remember that the two most popular logic families are TTL (Transistor-Transistor

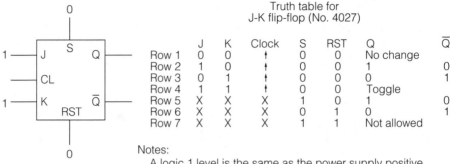

Truth table for
J-K flip-flop (No. 4027)

	J	K	Clock	S	RST	Q	Q̄
Row 1	0	0	↑	0	0	No change	
Row 2	1	0	↑	0	0	1	0
Row 3	0	1	↑	0	0	0	1
Row 4	1	1	↑	0	0	Toggle	
Row 5	X	X	X	1	0	1	0
Row 6	X	X	X	0	1	0	1
Row 7	X	X	X	1	1	Not allowed	

Notes:
A logic 1 level is the same as the power supply positive voltage.

A logic 0 level is the same as the power supply common connection ("ground").

An X means don't care. In other words, it doesn't matter whether it is logic 1 or a logic 0.

An arrow ↑ means a clock transition from low to high. In other words, the leading edge of the clock pulse.

■ **5-12** *J-K flip-flop with truth table.*

Logic) and CMOS (Complementary Metal Oxide Semiconductor) logic. A third family is ECL (Emitter-Coupled Logic), but, it is not popular in consumer products. Both TTL and CMOS logic families are used for making JK (and other types of) flip-flops.

The J-K flip-flop can be easily wired so that it will toggle (or divide by two) whenever a pulse is delivered to its clock input terminal. If the J-K flip-flop is made with TTL logic, it will usually toggle on the trailing edge of an input pulse to the clock input. If it is a CMOS flip-flop, it usually toggles on the leading edge of the clock pulse. Remember that those are not hard and fast rules! This is very important to know because you cannot mix TTL and CMOS flip-flops in a counting system.

While on that subject, you should never substitute any logic component from one family for a component from a different family—even though they have the same pinout. One reason in their power supply requirements for both voltage and current are likely to be different. Another reason is that they will surely have a different propagation delay. It will cause disastrous timing problems in an active circuit.

The truth table for a CMOS J-K flip-flop is shown in Fig. 5-12. An important thing to understand about J-K flip-flops is that with these devices, the SET and RESET terminals can be used to control the flip-flop regardless of the logic levels of J, K, or the clock.

The following discussion refers to the truth table for the NOR type, but, the terminology is similar for both types. Study each row of the NOR flip-flop truth table and make sure the meaning of each operation is clear.

In the first row there is no change with J, K, S, and RST at logic level 0. The RS section of this J-K flip-flop is in its stable condition when the SET(s) and RESET (RST) terminals are held to logic 0. In other words, J, K, and the clock control the output of this flip-flop. In the second and third rows, the outputs (Q and NOT Q) change with logic 1 levels delivered to either J or K. In the fourth row, the flip-flop toggles with the clock signal. In the fifth, sixth, and seventh rows, the S and RST inputs control the flip-flop regardless of the logic levels of J, K, or the clock.

Memories and microprocessors

Memories and microprocessors are being used in every major field of electronics. Unlike microprocessors in computers, dedicated microprocessors have their programs set by the manufacturer. Those programs cannot be changed.

Dedicated microprocessors are used extensively in consumer products. For example, microprocessors in VCRs and television receivers have made a significant change in the types of services being made available to the viewer. A very few examples are:

☐ The time of day can be displayed on the screen.

☐ The channel number can be displayed on the screen for a brief period when the channel selection is first made.

☐ The set can be programmed to automatically turn to the correct channel for selections made a week ahead of time.

☐ Nonmechanical pushbutton selections for the channels have been made available.

How does the microprocessor accomplish all of these things? In the technical literature, you see all types of strange characteristics assigned to the microprocessor. It is variously called the *brain* or the *heart* of a digital system. In reality, a microprocessor is so dumb that it is an absolute insult to compare it to a human brain or even the brain of a mouse. And, if anything should be called the heart of a system, it would be the clock (which all microprocessor and synchronous digital systems have). The clock is responsible for pulsing the data along the data busses and for regulating the continuing sequences of events that occur when the microprocessor is working.

To understand what the microprocessor really does, it is a good idea to go back and find out for what purpose it was originally designed. Originally, the microprocessor was designed to implement memory. Consider the plight of manufacturers of integrated circuit memory. From a relatively crude beginning of only a few memory locations, it very soon became possible to put 64,000 bits of memory on a single chip. Then, the real problems started. How do you get bits of information into and out of the memory without 64,000 wires?

The limit to the number of memories that could be sold was definitely based on the wiring complexity that the user or designer was willing to put up with. What was needed was a programmable device that would enable the user to get in and out of memory in some relatively simple way. That was why the microprocessor was first designed!

It is not uncommon to find literature today that refers to the microprocessor as a microcomputer. A purist would probably say that's an insult to the people that make computers. The reason people ascribe this feature to microprocessors is that one of the basic parts of a microprocessor is the arithmetic logic unit (ALU). This was put into the microprocessor so that data could be retrieved from the memory and operated on before it was presented to the outside world.

The ALU performs simple addition and subtraction and other basic arithmetic operations. It also performs the basic logic operations that are related to the seven basic gates, which were covered at the beginning of this chapter.

All of the microprocessors in use today can be divided into two basic categories: dedicated and undedicated. A *dedicated microprocessor* is programmed in the factory and can be used only for some particular application. The microprocessors used in television tuners, for example, are dedicated microprocessors. *Undedicated microprocessors* are very versatile, compared to the dedicated units in television tuners. The undedicated microprocessors can be programmed to perform an almost limitless number of tasks.

A second category for microprocessors is its number of bits. This is actually the number of bits of information that can be sent on the data bus at the same time. All microprocessor systems have at least three buses: the data bus, the address bus, and the control bus. The data bus carries the information to the memory and away from the memory. It also carries the information outside of the microprocessor and brings outside information in.

The number of bits of data on the data bus directly determines how many steps an operation must go through in order for a task to be completed. Some of the bits on the data line are used to identify the operation. If those bits aren't available, a separate step must be used. The tradeoff is in the fact that the greater the number of bits on the data bus, the more complex the programming operation, and the more expensive the microprocessor system.

In the television tuners and in other appliance applications, 4-bit microprocessors have been popular. Four-bit microprocessors were used because there just wasn't any need for any more bits. Well, at least that was the original contention. Because of the limited number of pins on a 4-bit microprocessor, there was some difficulty in getting all of the outside data into the chip.

The way the designers get around this problem is to use a multiplexer. A *multiplexer* is simply a logic circuit that permits a lot of inputs to be reduced to a single output. A *demultiplexer*, on the other hand, takes a single input and divides it up among a lot of outputs. Those multiplexers and demultiplexers make it possible to use operational microprocessors that have less pins available than the higher bit models. If you look at a microprocessor system, you will see that buses run from the microprocessor to the various memories (Fig. 5-13).

The random-access memory (RAM), which is a temporary storage device, is volatile. It is specifically designed to have the information stored in that memory changed frequency. The read-only memory (ROM) has the information stored during manufacturing. That information cannot be changed, and it is nonvolatile.

As microprocessor systems developed, a need emerged for read-only memories that were nonvolatile, but could be programmed in the field. In other words, the user did not want the memory programmed for eternity. That led to the development of the PROM (programmable read-only memory) and the EPROM (erasable programmable read-only memory). With this type, the original stored information could be erased with ultraviolet light and the memory could be reprogrammed in the field.

Using all of that equipment to reprogram the memory was inconvenient, so electrically erasable programmable read-only memories (EEPROMs) were developed. They are very popular in television systems. For example, when a customer moves into a

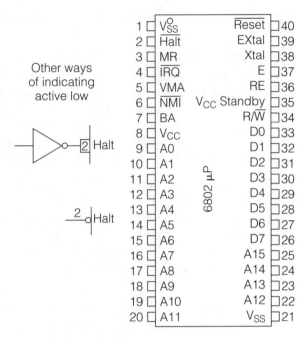

Other ways
of indicating
active low

1	V_{SS}^O	Reset	40
2	Halt	EXtal	39
3	MR	Xtal	38
4	$\overline{IRQ}$	E	37
5	VMA	RE	36
6	$\overline{NMI}$	V_{CC} Standby	35
7	BA	$R/\overline{W}$	34
8	V_{CC}	D0	33
9	A0	D1	32
10	A1	D2	31
11	A2	D3	30
12	A3	D4	29
13	A4	D5	28
14	A5	D6	27
15	A6	D7	26
16	A7	A15	25
17	A8	A14	24
18	A9	A13	23
19	A10	A12	22
20	A11	V_{SS}	21

6802 μP

■ **5-13** *Typical microprocessor.*

140

new area with his television set, he wants to program his TV receiver to receive the channels available. Once he has programmed it, he does not want to have to do that again (until he moves again). So, the read-only memory used to memorize the various channels was made to be electrically erasable.

If you are not comfortable with the theory and operation of microprocessors, you might find the authors' book (*Microprocessor Theory and Operation: A Self-Study Guide with Experiments,* Howard W. Sams, 1995) useful for a quick study.

An introduction to microprocessor (P) terminology

A good way to get familiar with microprocessor terminology is to review the names of pins of a typical microprocessor (Fig. 5-13).

A pin marked V_{CC} (pin 8) is a power supply connection to the microprocessor.

Two additional pins are marked V_{SS} (pins 1 and 21). They are the common or "ground" connections for the microprocessor. Why would a manufacturer put V_{SS} on both sides of the integrated circuit? The reason is simple. It simplifies the design printed circuit boards. A big problem in printed circuits boards is designing them so that conductors are not crossing over one another. It might be impossible to get the V_{SS} on one side of the microprocessor, but it is possible on the other side. So, the manufacturer has simplified the mounting of this device on printed circuit boards.

One pin is marked with a NOT HALT (pin 2). Any time you have a line over the top of a microprocessor or digital symbol, it means NOT or NO. What that means is that there must be a logic 1 into this pin to prevent the microprocessor from stopping what it is doing.

Two pins show input terminals for a crystal. In this microprocessor, the clock circuitry is built inside the package. Two terminals permit external crystal control of the clock circuitry.

Another pin is marked NOT INTERRUPT REQUEST (NOT IRQ). An interrupt request that comes from outside the microprocessor. It is a way of telling the microprocessor that there is information available that should be used and as soon as the microprocessor can get to it, it should service the request.

A pin marked VMA (Valid Memory Address) is very important. You must understand that memories cannot distinguish between data pulses and undesired pulses. The VMA pin is used by the microprocessor to tell the memory that this is really a signal; it is not fooling.

A NOT NONMASKABLE INTERRUPT (NOT NMI) pin tells the microprocessor that this interrupt must be taken seriously. The microprocessor cannot shut this interrupt out. It is an important interrupt signal to the microprocessor. By contrast, a plain interrupt tells the microprocessor that it has information to be serviced whenever it is convenient for the microprocessor.

The BA (Bus Available) pin acts like a streetcar conductor or a traffic cop. You wouldn't want to try to send signals into the microcomputer and out of the microprocessor both at the same time. The BA pin will prevent that.

The V_{CC} pin is for the positive 5 volts that is used for operating the microprocessor.

A total of 16 pins are marked with an *A* identification letter. The letter A represents addresses where information can be stored or sent. Because you can only use logic 1s and 0s for information, and you have 16 lines to carry the information, it follows mathematically that you have 2^{16} (65,536) different codes or addresses that you can identify on the 16 address bus lines. Another way of saying this is that this microprocessor can direct signals to 65,536 different address locations. The *address* is the place where information is stored in memory.

In microprocessor language, the number 65,536 is called *64K* (64,000). Usually a person studying this for the first time wants to know why they call it 64K, instead of 65,536. We don't really know the answer, but it probably goes back to the time when these types of memories were first being used. The manufacturer of the memories only guaranteed so many possibilities out of the total. In other words, if you were buying a memory that was supposed to be able to go to information at 65,536 addresses, you might only get 64,000. So, they used a lower number that they were willing to guarantee.

The pins marked with a *D* identification represent data. The microprocessor is involved with two different kinds of signals. One, called the *ADDRESS (A)*, shows where the information is going to or is coming from. The other, called the *DATA*, (D) carries the actual information in the form of a binary code.

Suppose you wanted to store the number 10_{10} in location 0000 0000 0000 0011_2. The first thing you would need would be the location where you want to store it. That is the...0011_2 address. The second thing you would need is the number or data. In this case the data to be stored is...1010_2.

An R/NOT W (READ/NOT WRITE) pin is used to operate certain types of memory. It tells the memory whether the μP (microprocessor) is putting information (data) into memory or taking information (data) out of the memory. In this sense, the memory is very much like your brain. Adding data is different than removing it. Most of us can't do both things at the same time, and neither can the microprocessor. The R/NOT W pin voltage prevents that from happening.

Pins marked (NC) on some microprocessors have no connections.

The pin marked DBE (Data Bus Enable) saves a lot of burned-up boards. It prevents the information from coming in and going out at the same time.

The pin marked TSC (Three-State Control) enables the microprocessor to make its data pins open circuited so that information can't be pushed in or out. The microprocessor is protected against signals being delivered to the data and address pins (and others) when the microprocessor isn't ready.

Suppose, as an example, that a data input is at logic 1; and, at the same time, that pin is at logic 0 as a result of an internal microprocessor action. Now you have 5 volts and 0 volts, both connected to the same pin. That is technically a short circuit and something has to give. The three-state control can make the pins open circuited and the short circuit is avoided.

The last pin to be discussed is marked *RESET*. When you first turn the microprocessor on, various internal sections of the microprocessor will be at 1s and 0s in a sort of random arrangement. It is necessary to clear all of those signals (or RESET) before using the microprocessor. That is what a signal on the RESET pin does.

Practice question

An 8-bit microprocessor is going to store 1110 0011 in the memory at 0000 0000 1110 0011. Which of the following is correct?

A. The address is 0000 0000 1110 0011 and the data is 1110 0011.
B. The data is 1100 0011 and the address is 1110 0011.

Answer: Choice A is correct. Notice that the data is 8 bits long and the address is 16 bits long. This is typical for an 8-bit microprocessor.

143

Other important IC logic systems and devices

Inside of the microprocessor is a necessity for very temporary storage of data while that data is waiting to be operated on by the ALU. This data is stored in a section called the *register* or, in some microprocessors, the *accumulator*.

Registers are also used in logic systems outside of the microprocessor. All registers can be divided into two basic classifications: parallel output data, in which all of the data goes out at the same time, or serial output data, in which the data goes out on a single (one-bit-at-a-time) line. Combinations of serial and parallel data in registers are possible (Fig. 5-14).

A moment's reflection will show you that a serial data system is necessary if you are going to send an 8-digit number over a telephone line. Because there is only one single telephone line, it is necessary to take the 8 digits and break them into a code that allows one digit to be sent at a time. So, serial output registers are used extensively for getting data in and out on telephone lines. Parallel registers are used extensively in microprocessors and other systems where the data can be delivered 8 bits at a time. That is assuming that it is an 8-bit system.

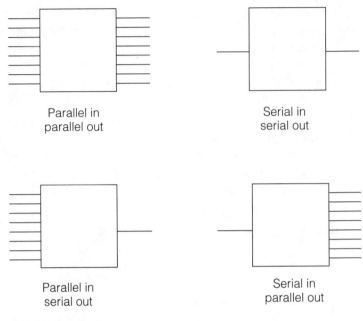

Parallel in
parallel out

Serial in
serial out

Parallel in
serial out

Serial in
parallel out

■ **5-14** *Types of registers.*

Figure 5-15 shows a microprocessor connected into a basic circuit. All of the individual sections shown in Fig. 5-15 might be built into a dedicated microprocessor.

If you were able to take apart a dedicated microprocessor from a tuner, you would find that it is not just simply a microprocessor. Most of those integrated circuits also have built-in RAMs and a ROM. To get all of that into a package, a very large number of pins is necessary on the chip. It is not uncommon to find 60 pins!

You might also find a phase-locked loop in the microprocessor IC. PLLs are used to tune a television receiver to a station and hold that station while the viewer is watching it.

Technicians report that ICs with many pins are difficult to get in and out of the circuit board because they have a tendency to break in half when you pry on one end! You are not even guaranteed against breakage by prying at both ends at the same time, but that is the least risky way to remove microprocessors from printed circuit boards. Of course, before you remove them, they usually have to be desoldered because it is not a common practice to use IC sockets.

In one of the CET tests, the term *bit slice* was mentioned and a large number of letters were received asking just what is a bit

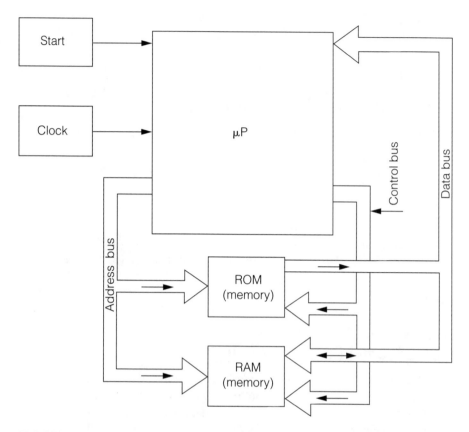

■ 5-15

slice. Well, if you look at the block diagram of an IC microprocessor, you will find that it consists of various individual sections (such as the ALU, the registers, the program counter, etc.). It should be obvious that those various sections could be located in individual IC packages rather than in a single package (like a microprocessor).

When you locate the various sections of a microprocessor system in separate packages, you have put together what is known as a *bit slice*.

What is the advantage of bit slicing? It is the fastest microprocessor system available today. The thing that makes a regular microprocessor slow compared to a bit slice is that the data has to be moved out and in to the various sections of the microprocessor system. A bit-slice design has no unnecessary sections, and only the absolutely necessary data motion is utilized in performing a desired function.

Bit-slice design has a great disadvantage in that it is much more difficult to program. You will generally not see bit-slice design in microprocessors used in television tuners. They are more likely to be found in industrial applications.

Additional terms you should know

Synchronous and asynchronous circuitry

A *synchronous circuit* requires the presence of another signal for its operation. The second signal is usually referred to as a *clock pulse*.

Glitches

Glitches are undesired signals of very short duration (such as transients). Glitches are produced in a circuit when the timing between two basic components is not exactly correct. Glitches have a very bad habit of destroying circuit operation.

Additional comments on TTL, CMOS, and ECL logic circuitry

The initials *TTL* represent *transistor-transistor logic*. This type of logic is very often numbered in the 7400 or 5400 series. The difference between these is in the tightness of the specifications, rather than in the pinouts or mode of operation. Thus, a 7400 is basically the same circuit as a 5400. Generally, the 5400 devices are designed for military operations with tight specifications, and the 7400 devices are used in commercial equipment.

One special branch of the TTL family contains the Schottky devices, which have an extremely fast switching time. As a matter of fact, it is possible to obtain switching times (propagation delays) of 3 nanoseconds per gate. That is not the fastest. Emitter-coupled logic gates are the absolute fastest of the IC families.

CMOS devices are made with combinations of N-channel and P-channel MOSFETs. They are often numbered in the 4000 series and can operate with voltages between 3 and 15 volts. They are not as fast as TTLs, and you will see typical propagation delays of 15 to 35 nanoseconds. Probably the most important advantage of CMOS devices is that they do not require a regulated 5-volt supply. CMOS devices have had a problem (especially in early CMOS integrated circuits) with static electricity. However, that problem is not serious in newer components, but be careful to avoid unnecessary exposure to static charges.

The abbreviation *ECL* represents *emitter-coupled logic*. This very fast logic family has propagation delays as little as 1.5 nanoseconds. An important characteristic of emitter-coupled logic is that they provide a constant drain on the power supply and therefore do not produce switching transients. Another important characteristic is that they are operated with a negative 5-volt power supply. ECL logic gates are normally numbered in the 10,000 series.

Programmable counters

A very important feature of many logic systems is a counter. Those for low-number counts (modulos) can be made with toggled flip-flops. Remember that either D or J-K flip-flops can be readily toggled.

For higher counts, manufacturers have put programmable counters into an IC package. They are characterized by the fact that they can count up to high numbers, or count down from high numbers. Also, they can be set to count up or count down to a desired value by applying the proper logic levels to the programmable inputs.

Synchronous and asynchronous counters

The terms *synchronous* and *asynchronous* are sometimes applied to counters. For the synchronous counter, all flip-flips change their state at the same time when a count is being made. That usually produces a rather heavy burden on the power supply. However, because the flip-flops all change at the same time, the synchronous counter is much faster than the asynchronous counter.

When an asynchronous counter adds a digit or subtracts a digit from the total count, it is necessary for the change to be produced by signals that ripple through each individual flip-flop. Thus, asynchronous counters are sometimes referred to as *ripple counters* or *ripple-through counters*.

The fact that not all of the flip-flops change their condition at the same instant means that asynchronous counters cause less drain is on the power supply. However, it takes longer for a count to be made because the flip-flops do not all change their condition at the same instant of time.

Encoders and decoders

Encoders and decoders are used to translate numbers from one digital system into another. An example of the use of an encoder would be to change the pressure on a typewriter key into the binary ASCII code, which can be recognized in a microprocessor system. A decoder example would be an IC that converts a binary count into signals for operating a seven-segment decimal number readout.

Static versus dynamic memory

A *static memory* holds a binary number over a period of time and does not require refresh signals. Flip-flops and latches can be used to produce static memories.

A *dynamic memory* requires a continuous recirculating refresh signal to maintain the memory count. Dynamic memories are sometimes made by charging capacitors. The charged capacitor can be used to indicate that a binary 1 is present and the uncharged capacitor can represent a binary 0. Once a capacitor is charged, it has to be continually recharged, or refreshed, to maintain the memory count. Special integrated circuits have been designed to do that.

In comparing dynamic and static memories, understand that dynamic memories are much faster for access and it is possible to put a very large number of memory cells on a single IC. The disadvantage is that the refreshed circuit requires a constant recharging energy.

Charge-coupled devices (CCDs)

Charge coupling occurs when static charges are stored under the surface of certain materials. These charges can represent 1s and 0s, depending on whether or not they are present. The characteristic of charge coupling that is most important is the high density (number of memory cells per chip) and the very rapid speed with which the information can be entered and retrieved.

A/D and D/A converters

Analog-to-digital and digital-to-analog converters are necessary because much of the electrical information available in the outside world is not suitable for operating microprocessor and digital systems. Therefore, the data has to be converted from an analog (linear) format into a digital format so that it can be utilized in the microprocessor and digital systems. Conversely, if the information is to be sent to the outside world, the digital information might not be suitable for operating linear systems, so the digital signal must be converted from digital to analog.

Please refer to Appendix C for practice questions on the subjects in this chapter.

Television and VCRs

6

EXPERIENCED JOURNEYMAN TELEVISION TECHNICIANS DO not normally have trouble answering questions in this section of the CET test. Monochrome and color television circuitry are covered in the television section of the test. You might be asked questions about the television signal and that information is included in this review.

A separate section in the CET test has questions on videocassette recording and videodisc recording. Material for that section is also reviewed in this chapter.

Television and FM signals

The distribution of frequencies in a television *IF (intermediate frequency)* stage are shown in Fig. 6-1. A 1.25-MHz vestigial sideband is transmitted as part of the *RF (radio frequency)* signal. This vestigial sideband and the carrier are included with the transmitted signal to simplify tuning the television receiver. Neither carries any useful intelligence as far as reproducing the picture or sound is concerned.

The 4.5-MHz frequency range between the picture carrier and the sound carrier is very important. In a monochrome receiver, it is not uncommon to use a product detector to separate the sound signal. In that circuit, the picture carrier is heterodyned with the sound signal to produce a sound IF signal. That is not done in a color TV receiver. Instead the sound signal IF must be taken off at some point before the detector in order to avoid heterodyning between the sound IF and the 3.58-MHz chroma signals.

Figure 6-2 shows the relative frequencies of the oscillator, video carrier, and sound carrier frequencies as they are delivered to the tuner. The carrier and oscillator frequencies happen to be for a particular VHF channel. Notice that the difference in frequency between the local oscillator and sound carrier is less than the difference between the local oscillator and the video signal carrier.

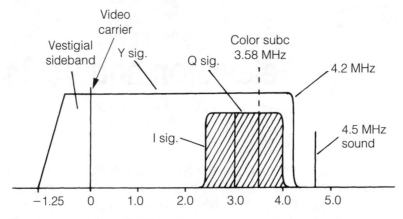

6-1 *Distribution in TV IF.*

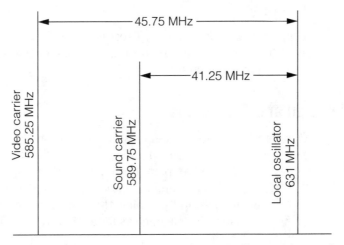

6-2 *Reductive frequencies of the oscillator video and sound tuner.*

After the three frequencies in Fig. 6-2 are heterodyned in the tuner, the sound signal will have a lower frequency (41.25 MHz) than the video signal (45.75 MHz). The local oscillator frequency is higher than the sound and video signal frequencies.

Figure 6-3 shows the distribution of television signals for channel 33 after they pass through the tuner. As we mentioned, the video carrier frequency (at 45.75 MHz) is above the sound carrier in frequency (41.25 MHz). That is opposite to the condition shown in Figs. 6-1 and 6-2.

Figure 6-4 shows the distribution of relative signal amplitudes for video blanking pedestals and synchronizing pulses. The signal is

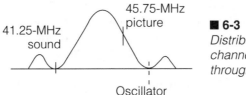

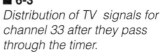

■ 6-3

Distribution of TV signals for channel 33 after they pass through the timer.

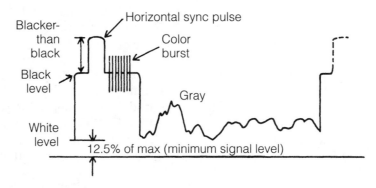

■ 6-4 *Relative signal amplitudes.*

never allowed to go below 12.5 percent of maximum, so 100% modulation is never achieved with a television signal.

Referring to Fig. 6-4, about eight cycles of color burst are located on the "back porch" of the horizontal synchronizing and blanking pedestal.

A minimum of eight cycles is transmitted for this burst. This burst signal is the original color carrier of the transmitter system. As you know, the color carrier is suppressed and only sideband information is sent for each of the color signals. Therefore, some form of carrier reinsertion is necessary in the receiver.

The color burst offers the reference that is necessary for the color carrier to be reinserted in the proper frequency and phase relationship that it had at the transmitter. In the color section of the receiver, the burst controls the frequency and phase of the color oscillator.

Two important signals are sent during the blanking period between television fields. You will remember that two fields are required to make one frame (complete picture). Each field is actually a half picture that has been obtained by scanning the odd lines for one field and the even lines for the next field.

The blanking period between the two fields is transmitted as a blanking pedestal. In the actual picture, it shows up as a dark space, sometimes called a *blanking bar*, if you roll the picture vertically to a half of a frame.

The VITS signal (Vertical Interval Test Signal), which is visible during vertical blanking, is used primarily to evaluate the quality of television signals. Experienced journeyman CET technicians have indicated to us that they are using this VITS signal as one method of evaluating receiver performance.

A second signal that is sent during the vertical blanking period is the VIRS (Vertical Interval Reference Signal). This signal is used in some receivers to automatically adjust the color of the picture delivered by the receiver. By that method the receiver color closely matches the color of the original scene at the transmitter. Of course, special VIRS signal processing is required for that feature.

In a monochrome system, the horizontal sweep frequency is generally given at 15,750 Hz. That frequency had to be modified slightly to 15,734+ Hz when the color signal was added into the NTSC monochrome signal. However, 15,750 Hz is still generally given as the horizontal oscillator frequency.

The equation relating period and frequency is:

$$T = \frac{1}{f}$$

where: T is the time for one cycle
f is the frequency of the signal

The equation can also be used to find the time for the horizontal sweep:

$$T = \frac{1}{15,750} = 63.4 \text{ microseconds}$$

where: T is the time between horizontal sync pulses on the transmitted signal

It is often designated H and used as a time reference for all TV signals. That includes the times for the blanking pedestal, sync pulse, and one line of video.

Every second, 60 fields are transmitted, so the time for one field is also easily computed as:

$$T = \frac{1}{60} = 16,667 \text{ milliseconds}$$

It takes two fields to make one frame. Therefore, the time for one complete frame is equal to twice the time for one field: time for frame, $T = 2 \times 16.667$ ms $= 33.333$ milliseconds.

Color TV

The color television picture is made up of three color primaries: red, green, and blue. These are additive primaries. They are not to be confused with the type of primary colors that you used when you were playing with crayons. Those were *subtractive primaries*. With subtractive primaries, the colors yellow and blue combine to make green. With *additive primaries*, the colors green and blue combine to make the color cyan.

Of special importance is Illuminant C, the equivalent of daylight. To understand why this is important, consider what would happen if you added the red, green, and blue colors in equal amounts. The result would be a white with a very heavy blue tint, which would be uncomfortable to watch.

When you are watching a color picture of a daylight scene, the sunlit areas should match (as closely as possible) the sunlit areas in your real-life experience. To accomplish this, the three primary colors must be added in the following proportions:

Illuminant C = 11% blue + 59% green + 30% red

FM

As a consumer electronics technician, you should be familiar with the makeup of the FM broadcast multiplex signal (Fig. 6-5). All of the frequencies shown in Fig. 6-5 are with reference to the FM carrier frequency. For example, the pilot frequency is 19 kHz above the FM carrier frequency.

The 19-kHz pilot is used only as reference so that the 38-kHz signal carrier can be reinserted. In the receiver, this pilot signal is doubled in frequency and used for carrier reinsertion to demodulate the L-R signal.

The 67-kHz SCA subcarrier can be used to transmit music without accompanying advertisements. At the FM radio transmitter, an SCA decoder would be used to play music in places such as a physician's waiting room. The SCA subcarrier can also be used to transmit data or other information.

153

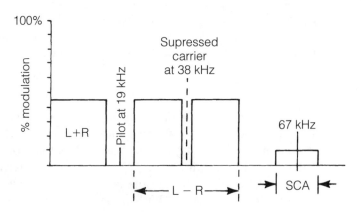

■ 6-5 *The FM broadcast signal frequency distribution for one station.*

The television receiver

You should be familiar with the monochrome and color receivers in block diagram form. Figure 6-6 shows the block diagram of a monochrome receiver.

In the CET test, you might be shown a block diagram like this and be asked to identify the names of the blocks in various parts of the receiver. Figure 6-7 is a highly generalized color TV receiver that does not represent any particular brand. As with the monochrome

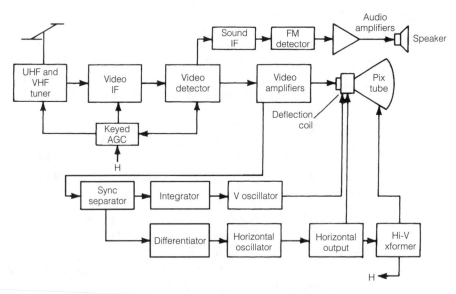

■ 6-6 *Block diagram of a monochrome receiver.*

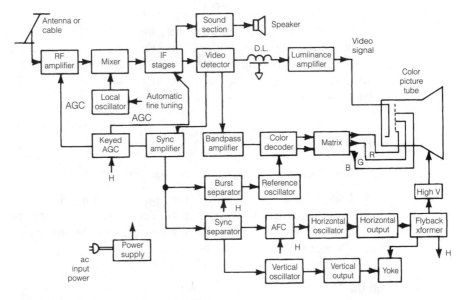

■ **6-7** *Block diagram of a color TV receiver.*

block diagram, you might be asked to identify blocks in this type of drawing. An output pulse from the flyback transformer is used by the keyed *AGC (Automatic Gain Control)* and *horizontal AFC (Automatic Frequency Control)*. It is also used in the burst separator to get the burst off the back porch of the horizontal blanking pedestals.

In color receivers, the horizontal lines are made digitally from the transmitted signal. The incoming signal is delivered to the UHF and VHF tuner and RF amplifier where it is mixed with the local-oscillator signal in a stage called the *mixer* to produce the IF signal.

In radio receivers, the RF stage is often eliminated and the mixing takes place in a converter. The mixer output IF signal is amplified in several stages. Just before the video detector, the sound signal is separated to be processed in a different section of the receiver.

In the particular block diagram, the output of the video detector goes to three different places. One is to the delay-line (D.L.) luminance amplifier. Think of it as a video amplifier with a different name.

The second output from the video detector goes to the color bandpass amplifier. Its purpose is to eliminate most of the luminance signal and pass only the *I* and *Q* color signals.

After those signals have been amplified, they are decoded and delivered to a matrix (color decoder) for distribution to the three color grids in the color picture tube. In some receivers, the luminance signal is combined in a matrix to produce the red, green, and blue signals. In another type of receiver, these signals are combined inside of the television picture tube.

Practice question

Why is a delay line needed in the luminance section?

Answer: The color signals go through more amplifiers, so they would arrive late at the picture tube. As a result, the color and luminance signals would be out of registration if the delay line did not slow down the luminance signal.

Sync and burst

The third output of the detector goes to the sync amplifier. It produces synchronizing signals for the sync separator and also for the burst separator. Remember that the burst sits on the back porch of the horizontal blanking pedestal. During color programs the burst is removed by a coincidence circuit that has an input pulse from the horizontal flyback transformer. When the horizontal pulse arrives, it permits the burst amplifier to amplify the burst on the back porch. At all other times during the sync signal, the H input is not present, and there is no output from that amplifier.

The burst signal is used to synchronize the reference oscillator, then the oscillator signal is recombined with the I and Q quadrature signals to produce signals, which are similar to the ones in the transmitter before the carrier was removed.

Sync separator and deflection circuits

The sync separator produces an output signal in both the horizontal and vertical oscillator sections. Although a differentiating circuit is usually used to retrieve the horizontal sync signals, an integrating circuit is used to retrieve the vertical sync signals.

In the horizontal section, an automatic frequency control (AFC) in the horizontal module holds the horizontal oscillator to the correct frequency—even though noise might be present in this section. The output of the horizontal oscillator goes to a horizontal output amplifier, then to the horizontal output transformer (HOT).

In newer receivers, digital circuitry is used to retrieve horizontal and vertical sync pulses. The result is a more stable picture that is less likely to be disturbed by noise.

The flyback transformer produces the high voltage that is necessary for the dc second anode and focus voltages on the picture tube. It does this with the help of the voltage tripler. The flyback transformer is also the source of the signal that is used for various processing in the color receiver.

The vertical oscillator and vertical output signal, as well as the horizontal signals, are delivered to the deflection yoke. The H and V output signals move the beam back-and-forth and up-and-down on the screen. The rectangular area on the screen that is swept by the deflected beam is called the *raster*.

The AFC circuit

Although not shown in the illustration, an *AFC (Automatic Frequency Control)* circuit might be present in the tuner. It gets its signal from the IF stage and is used to lock the local oscillator signal onto the frequency that is required to demodulate the carrier.

Ideally, the local oscillator is always an exact frequency above the carrier frequency of the incoming signal. If the oscillator drifts, the color picture portion of the signal can very quickly be downgraded or lost. So, this AFC circuit (also called *AFT, automatic fine tuning*) is designed primarily to maintain color television reception. In later sets, a *PLL (Phase Locked Loop)* is used to lock the local oscillator onto the proper frequency.

AGC circuit

The automatic gain control (AGC) in television receivers is usually a keyed type. The strength of the AGC signal depends on the steady strength of the sync pulses on the incoming signal.

It would be difficult, if not impossible, to get an AGC voltage off of the video signal because the picture changes continuously with the brightness and darkness as the scenes change. Using the sync pulses as a reference is more logical because their amplitude is directly related to signal strength. Obtaining a dc voltage that is directly related to the signal strength is accomplished in the keyed AGC circuit.

Note the H (Horizontal sync pulse) output from the flyback transformer and the H input to the AGC section in the block diagram of Fig. 6-7. That H pulse occurs at the same time as the horizontal sync pulse. So, it is used to gate the AGC circuit ON only during the time when the horizontal sync pulses are present. In other words, the AGC amplifier is keyed on by the 'H' signal. Therefore, the output of the keyed AGC circuit sync amplifier consists only of the sync pulses. Those sync pulses are rectified to produce a dc voltage that controls the gain of the RF and IF amplifiers in the system.

Power supply

The power supply in the block diagram of Fig. 6-7 converts the ac input power into dc voltages that are necessary to operate the various sections of the receiver. A disadvantage of this type of system is that all the current comes through the 60-Hz power transformer. The power transformer is very heavy and very costly.

A more modern system uses what is known as a *scan-derived power supply*. In that system, the dc voltages for operating the various stages are obtained by rectifying the scanning voltage output from the flyback transformer.

With that type of scan-derived power supply circuit, it is necessary to have a startup circuit. In other words, you won't get any voltage out of the flyback transformer until the horizontal oscillator and horizontal output stages are working. At the same time, the horizontal oscillator and horizontal output stages won't work if there is no output from the flyback transformer.

So, a starting circuit is necessary to get the circuit into operation. Once the circuit is started, the startup circuit is automatically disconnected. Then, the circuit operates primarily from the scan-derived voltages that are obtained from the flyback transformer.

Comb filter

Figure 6-8 shows a block diagram of a somewhat different color television receiver design. The tuner and the IF module are similar to those in the block diagram of Fig. 6-7. Likewise, the sound module, which receives its signal from the IF stage before the detector, is also similar to the one previously covered.

The video signal is detected and the output video signal is delivered to a comb-filter module. The comb filter is designed to separate the color signal from the luminance (Y or brightness) signal.

As you can see in the television passband drawing of Fig. 6-1, the color and luminance signals exist in the same region of the transmitted signal. Special circuitry must be used to keep each of these signals separated in the receiver.

158

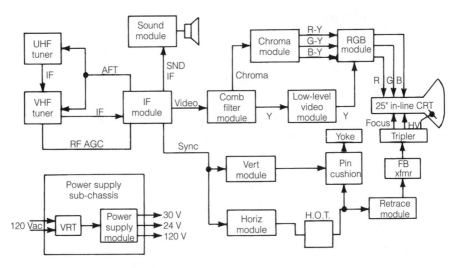

6-8 *Power supply block diagram and a different color TV receiver.*

One way to do that is to use filters in the bandpass amplifier to keep out the luminance signal. Likewise, filters can be used in the luminance section to keep out the color signal. However, a comb filter does that much more efficiently. It is a very sharp cutoff filter that eliminates interference between the color and luminance portions of the signal.

Observe that the comb filter has two outputs. One goes to the chroma module where the color signal is demodulated to produce the red, green, and blue signal voltages. The other produces the Y (luminance signal). That signal also goes to the matrix. In this particular receiver, the luminance and the color signals are combined in the RGB module. Then, they are delivered to the picture tube.

Pincushion and color killer

The remainder of the color receiver in Fig. 6-8 is similar to the one covered with reference to the previous block diagram. Notice that pincushion correction is obtained in a pincushion module. This module produces signals that prevent sagging in the vertical and horizontal perimeter of the raster.

Not shown in the block diagrams is the color-killer circuit. Its purpose is to shut off the color amplifiers when no color signal is received. This circuit is important because it eliminates the so-called "colored confetti" (which is colored snow) produced by amplifiers in the color section. Colored confetti is generated by the bandpass amplifier in the absence of an incoming color signal. Remember,

159

the color killer output is off when the bandpass amplifier is passing a signal. The color killer comes on to shut the amplifier off during a monochrome signal (the absence of a color signal).

Practice question

Which of the following is a more likely cause of colored confetti in a color TV black and white (monochrome) picture?

A. Loss of burst
B. Color killer is not on

Answer: Choice B is correct.

Practice question

Refer to Fig. 6-9. What is the name of the section marked with an "X"?

Answer: IF

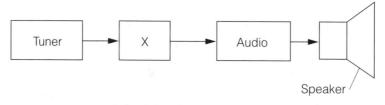

■ 6-9

Video recorders

We are very grateful to NAP (North American Philips Company) for supplying the following material on video cassette recorders and players. Video tape recording equipment has been used for many years by the military, television broadcasters, educational institutions, and other specialized industries. With the advancement of technology, the performance of these machines has reached a very high level.

The metallic-coated mylar tape on which the program information is recorded is housed in a cassette cartridge. The machine design is based on the Video Home System (VHS) standards so programs recorded on one machine can be played on a machine of another brand, which has been designed to the VHS standards.

To make a video tape recording, the picture intelligence is changed into electrical signals, which produce magnetic fields to magnetize the metallic coating on the tape. The device that changes the electrical signals to magnetic fields is called the *video head*. Most of us are aware that audio tape recorders also use a "head" to convert the electrical signals to magnetic fields. However, the video head design is quite different than the audio head design. In fact, the video heads used in the VHS system represent a technical breakthrough that allows a consumer-priced video tape recorder to be manufactured and marketed.

The most important difference between video and audio tape recording is the frequencies involved. The audio frequency range is about 20 Hz to 20,000 Hz. The video frequencies, however, range from 30 Hz to 4 MHz, as shown in Fig. 6-10. As a result, the VHS machine uses two different type heads for these purposes. One type for audio and another type for video. The audio head is similar to that used in an audio cassette tape recorder, but the video head is different because of the high frequencies involved.

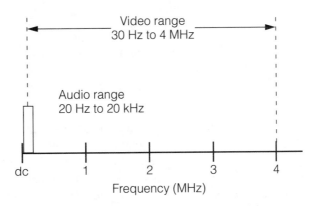

■ **6-10** *Range of video frequencies.* (NAP)

Figure 6-11 shows the basic construction of a tape head. A core of some metallic material has a coil of insulated wire around it. The core has an air gap on one side. This air gap contacts the tape.

Electrical signals applied to the coil of wire produce magnetic lines of force that travel through the core. When the metallic coating of the tape is against the air gap, the magnetic lines of force travel through the tape metallization to get to the other side of the gap. In so doing, the tape becomes magnetized according to the signal

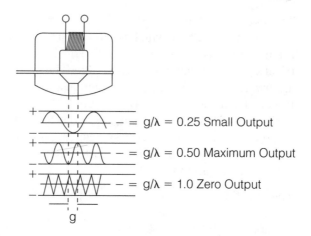

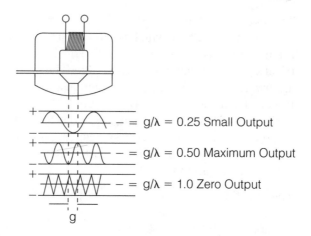

g/λ = 0.25 Small Output

g/λ = 0.50 Maximum Output

g/λ = 1.0 Zero Output

g

■ **6-11** *Construction of tape head.* (NAP)

that passed through it. During playback, the head acts as a pickup and the magnetized tape produces a magnetic field in the core as the tape moves by the air gap. This magnetic field induces an electrical signal in the coil of wire.

The width of the head air gap is very important and is the main reason why an audio head cannot be used with video frequencies. The head gap determines the highest frequency that can be handled by the head.

Figure 6-11 shows the wavelength on the tape of three different frequencies compared to head gap. The smaller the potential difference within the head gap, the smaller will be the output signal. Notice that the maximum output from the head is achieved when the gap equals one half the wavelength (gap-to-wavelength ratio = 5). When the frequency gets so high that its wavelength equals the gap, no output is achieved (gap-to-wavelength ratio = 1).

It is important to realize that wavelength, as used here, means wavelength on the tape. Therefore, a fast tape speed lengthens the wavelength of a given signal, but a slower tape speed shortens it. This point is illustrated by the following formula:

$$\lambda = V/F$$

where: λ = recording wavelength (in inches)
 V = tape speed (in inches/sec.)
 F = frequency of signal (in cycles/sec.)

Thus, higher frequency signals can be recorded by increasing tape speed and decreasing the head gap. However, practical limits are involved. A fast tape speed means much tape and large reels are required. Also, heads with tiny air gaps cannot be constructed easily.

Take some examples: If a tape speed of 15 in/sec and a frequency response to 20,000 Hz is desired, then according to the above formula, the wavelength on the tape would be: 15/20,000 = 0.00075 = 0.75 mil (75 thousandths of an inch). A head gap in this case of 0.75 mils would produce no output at 20,000 Hz, but a head gap of half that size (0.375 mils) would produce maximum output at that frequency. A head gap of 0.3 mils is commonly used in audio tape recorders.

If the same principle is applied to a 3-MHz video signal at 15 in/sec, a head gap of 0.0025 mils is needed. If a head gap of 0.3 mils were used, the tape speed could be increased to allow 3-MHz recording, but 900 inches per second would be consumed.

One way to get around this problem is to increase the relative speed between the tape and the head by moving the head and the tape at the same time instead of just moving the tape past a stationary head. The VHS machine does just that. The video head is placed on a rotating cylinder which spins at 1800 RPM while the tape slowly moves by. Combining this technique with a narrow head gap of 0.01378 mils (0.35 microns) allows the necessary head-to-tape speed while providing desirable lengths of recording time with minimum tape consumption.

An additional obstacle to video recording is the wide frequency range. The video spectrum ranges from 30 Hz to 4 MHz. This range represents about 18 octaves (when a frequency is doubled, it is said to have increased one octave). The reason this wide range is an obstacle is shown in Fig. 6-12. The head output volt-

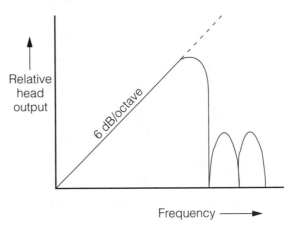

■ **6-12** *Wavelengths vs. head gap.* (NAP)

age level is very low for low frequencies and increases to its maximum of 100 dB at a 6 dB/octave rate. The difference in head output between low frequencies and high frequencies is 100 dB; much too wide in frequency for practical head design to accommodate. Incidentally, the two peaks at the right side of the graph are for higher frequencies beyond the video spectrum, whose wavelengths happen to be in multiples of twice the head gap.

A good way to solve the excessive output range problem is to move the required 4-MHz bandwidth up the frequency spectrum. For example, 20 Hz to 4 MHz is 18 octaves. 4 MHz to 8 MHz is still a 4-MHz bandwidth, but it only represents one octave. Thus, only a 6-dB difference in output level exists between low and high frequencies. This amount of level difference can be handled through equalization.

As shown in Fig. 6-13, the VHS machine actually uses a 3.4-MHz carrier, which is frequency modulated with the composite video signal. Sync tips are at 3.4 MHz while peak whites deviate the carrier to 4.4 MHz. The FM spectrum is from about 2 MHz to 6 MHz including sidebands. Also note that the chroma is converted down from 3.58 MHz to 629 kHz in order to prevent beat products because of 3.5-MHz heterodyning with the FM carrier.

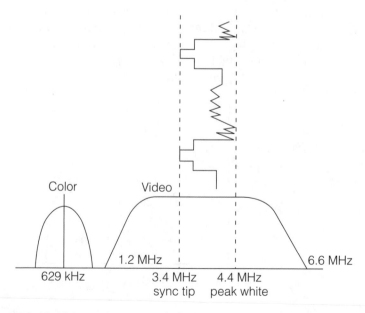

■ **6-13** *Video spectrum.* (NAP)

VHS principles

The VHS system uses several unique principles. Among these is a recording technique known as *helical scanning*. Figure 6-14 illustrates the concept. Two video heads are used, A and B. They are attached to a tilted cylinder rotating at 1800 RPM. Although the cylinder is tilted, the tape remains on a horizontal path. As a result, each head scans the tape at an angle from bottom to top. The tape is wrapped 180 degrees around the cylinder and the heads are located 180 degrees apart. Therefore, only one head touches the tape at a time.

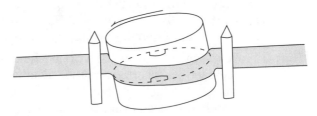

■ **6-14** *FM modulation of video.* (NAP)

Figure 6-15 shows the results of helical scanning with two heads. As each head moves across the tape it is recording one TV field (262½ lines). One revolution of the head cylinder allows each head to cross the tape one time, thus one revolution records two fields or one TV frame. Helical scanning is required in order to make each video tape (the path traced by a head) long enough to include 262½ lines (approximately 3¾").

Figure 6-16 shows the ½" tape and the arrangement used during recording. The audio track is across the top of the tape while the

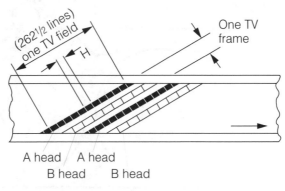

■ **6-15** *The concept of helical scanning.* (NAP)

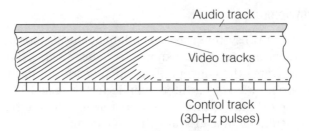

Audio track

Video tracks

Control track
(30-Hz pulses)

■ **6-16** *The result of helical scanning with two heads.* (NAP)

video tracks are recorded in the center of the tape. The bottom track is called the *control track*. The 30-Hz control pulses, which are used for synchronization during playback, are recorded here.

The path followed by the tape in the VHS machine is shown in Fig. 6-17. When a cassette is inserted and the machine is set to play or record, the slanted loading pins pull the tape out of the cassette and wrap it around the head cylinder, as shown. Posts P2 through P4 guide the tape keeping it at the correct height in the horizontal plane.

The tape moves from left to right. As it does, the tension arm provides back tension to the tape. The full erase head has a 67-kHz

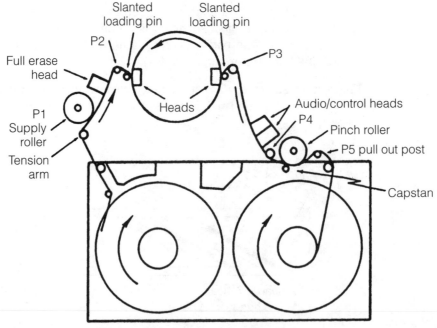

■ **6-17** *Recording on ½" tape.* (NAP)

signal applied to it, which fully erases the entire tape during the record mode. The Tilted Loading pins angle the tape so it enters and leaves the head cylinder parallel to the tilted head cylinder. After the tape has passed around the head cylinder, it goes past the audio/control head assembly.

This assembly actually contains three heads: one for audio erasure which is used for audio dubbing, one for audio recording or playback, and one for recording or playing back the 30-Hz pulses on the control track. Also present is the pinch roller, which presses the tape against the moving capstan for tape drive.

Overall block diagram

The diagram in Fig. 6-18 shows an overall view of the VHS portable system. It is divided into two sections: the deck unit and the tuner unit. The tuner has two major sections, the tuner/demodulator and the power supply. The deck unit has four major sections: audio, video, servo, and system control. Video signal processing includes the record and playback electronics. Servo control includes drive and control of the four direct drive motors, which are the cylinder, capstan, supply reel, and takeup reel motors.

Both UHF and VHF antennas are connected to the machine. From the VCR, the are routed back to the TV set. In this manner, the TV can be viewed normally when the VCR is not in use. Notice, however, that the TV/VCR switch must be in the "TV" position to route the VHF antenna back to the TV set. In the "VCR" position, the output of the RF converter (channel 3 or 4) is applied to the VHF antenna of the TV set.

The VCR has its own tuners. The desired channel to be recorded is selected by the VCR tuners. A complete TV demodulator is included (IF, AGC, AFT, sound) in the VCR.

The audio and composite video signals at the output of the tuner are coupled through the normally closed contacts of the 10-pin camera jack and the external input jacks to the appropriate section. If an audio or video line signal from an external source (i.e., video camera or another VCR) is connected, the output of the tuner unit is automatically disconnected and the external signals are coupled into the signal processing section.

The audio signal is applied to the audio circuits where it is equalized and amplified before being applied to the audio head. A 67-kHz bias current from a bias oscillator is coupled to the audio circuits for proper head bias as well as to the audio and full erase heads for tape erasure during the record mode.

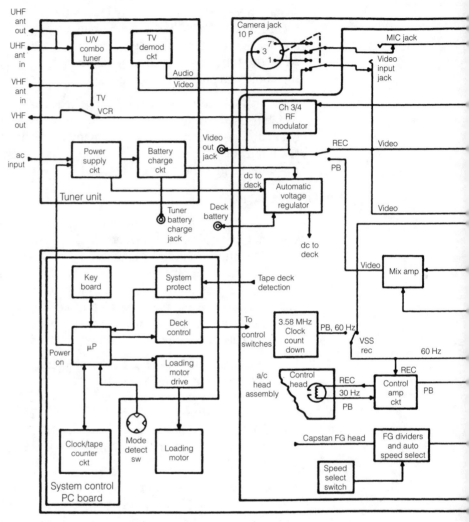

■ 6-18 *Overall block diagram.* (NAP)

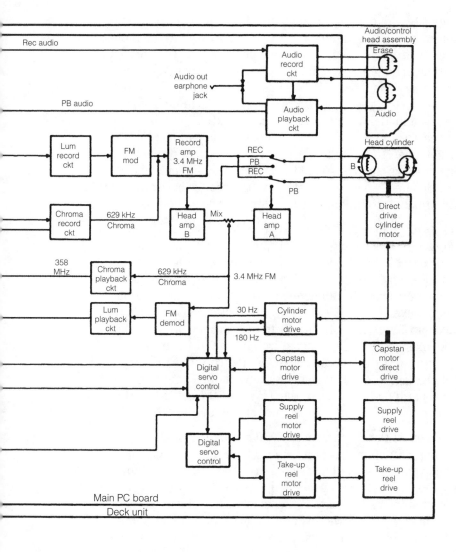

Rec audio

Audio/control
head assembly

Audio
record
ckt

Erase

Audio out
earphone
jack

PB audio

Audio
playback
ckt

Audio

Head cylinder

Lum
record
ckt

FM
mod

Record
amp
3.4 MHz
FM

REC
PB
REC

PB

B

Chroma
record
ckt

629 kHz
Chroma

Head
amp
B

Mix

Head
amp
A

Direct
drive
cylinder
motor

358
MHz

Chroma
playback
ckt

629 kHz
Chroma

3.4 MHz FM

Lum
playback
ckt

FM
demod

30 Hz

Cylinder
motor
drive

180 Hz

Digital
servo
control

Capstan
motor
drive

Capstan
motor
direct
drive

Supply
reel
motor
drive

Supply
reel
drive

Digital
servo
control

Take-up
reel
motor
drive

Take-up
reel
drive

Main PC board

Deck unit

169

The composite video signal is split into luminance and chrominance information. The luminance signal is processed in the luminance record circuit while the chroma signal is sent to the chroma record circuit. The luminance information frequency modulates a 3.4-MHz carrier. The 3.58-MHz chroma is down-converted to 629 kHz and applied to a record amp along with the FM luminance signal. The combined FM luminance plus 629-kHz chroma signal from the record amp is sent to both video heads in parallel. No bias current is applied to the video heads. The FM signal requires no bias, but serves as the bias signal for the 629-kHz AM chroma signal. The video heads, located on the rotating head cylinder, transfer the information to the tape as it is drawn past the heads.

In the playback mode, each video head couples its signal picked up from the tape to a head amp. The outputs of head amps A and B are added together and applied to the luminance processing and chroma processing circuits. Luminance processing demodulates the FM and retrieves the luminance video signal, while chroma processing converts the 629-kHz chroma back to 3.58 MHz. The video mix amp combines both signals to recreate the original composite video signal. The final signal is sent to the RF modulator along with the retrieved audio signal to modulate the RF carrier for either channel 3 or 4. This signal is then suitable for application to the VHF antenna terminals of the TV set. If the set is tuned to channel 3 or 4, it will receive the picture and sound from the VCR.

The remainder of the drawing deals with servo control. A *servo* is a closed loop amplifier that is used for electronic control of mechanical movement. A feedback signal from the mechanical movement to the electronic control completes the servo loop and maintains its stability. In this case, the mechanical movement to be controlled in the VCR is the cylinder motor, the capstan motor, and both reel table motors. The cylinder motor spins the upper cylinder on which are located the two video heads, while the capstan motor provides constant tape movement past the video heads.

It is necessary to rotate the Cylinder Motor at 1800 RPM so that one TV frame is recorded on the tape with each revolution of the Head Cylinder. Thus, each head will record one vertical field (262½ lines) on each track of the tape. In addition, it is necessary to place vertical sync at the same spot on every track (about 10H up from the bottom of the tape track). In order to maintain both of these criteria, both speed and phase servo loop circuits must be involved in the control of the cylinder motor. The feedback signal for the speed loop is a 180-Hz square wave. The square wave can be considered an FG (frequency generator) signal because its frequency is dependent on the RPM of the head cylinder. This signal is derived from the main coil drive signal. The phase loop feedback is a 30-Hz head position pulse generator (PG) signal.

Two magnets are mounted on the flywheel in near perfect alignment with the heads on the head cylinder. One magnet is polarized N-S while the other is S-N. These magnets will pass over a stationary hall IC pickup as the flywheel spins, producing both negative (A head) and positive (B head) PG pulses. This pulse signal indicates that the video head is at the bottom of the tape ready to record another field. Hence, it would seem that if we would synchronize the 30-Hz vertical field rate of the video signal during recording with the 30-Hz PG pulse, then each video head would start recording on the bottom of the tape at the beginning of each video field. During playback, the same setup cannot be used because there can not even be a video input connected. Instead, the 60-Hz reference is provided by dividing a 3.58-MHz crystal oscillator by 59.712.

The 30-Hz PG pulse is also used to indicate which video head is contacting the tape, A or B. This is important because, during playback, it is necessary to turn the Head off when it is not scanning the tape to prevent pickup of extraneous noise.

During the record mode, the 60-Hz vertical sync from the vertical sync separator in the chroma circuit is divided to 30 Hz in the control amp circuit. These 30-Hz pulses are applied to the control head and recorded on the control track of the tape. During playback, these 30-Hz control pulses are applied to the capstan motor servo circuits and used as the feedback signal for the capstan phase servo loop. Because the 30-Hz control pulses we derived from vertical sync, any head-to-tape track alignment variations can be compensated by slightly advancing or retarding the tape movement. This ensures perfect head-to-tape track alignment and results in maximum signal picked up off the tape.

Approximate capstan motor speed is maintained by feedback from the speed sensor frequency generator (FG). In the SP mode, the capstan FG is 1080 Hz. In LP, the tape moves at one-half of the SP speed; thus, the FG is 540 Hz. In SLP, the tape speed is reduced to one-third so the FG is 360 Hz. The SP/LP/SLP divider circuit divides each FG signal so that the FG signal sent to the capstan motor speed servo circuit is 270 Hz in SP and LP or 360 Hz in SLP. If the capstan motor should drift in speed, the capstan FG frequency will change and the motor drive circuit will correct the speed of the motor. During recording, the capstan FG is divided to create a 30-Hz pulse to be used by the phase servo circuit.

The SP/LP/SLP select circuit determines at what speed the tape is recorded or played back. In the record mode, the speed select circuit reacts to the input from the speed select switch and causes the motor servo and divider circuits to speed up or slow down the tape travel, as indicated. However, during playback, the capstan motor speed is automatically shifted whenever a large change in the frequency of the control pulses is sensed (i.e., LP/SLP, 30 to 90 Hz).

Overall block diagram troubleshooting

Check each of the following in order:

1. Power on
2. TV/VCR switch
3. Timer switch
4. DEW indicator
5. Playback only (by playing a pre-recorded tape)
   A. Head output
   B. Luminance and chroma outputs
   C. RF converter input and output (by switching TV/VCR switch and using tuner)
6. Record only (make a recording and play on different VCR)
   A. Camera/tuner switch
   B. Tuner (apply a video or audio signal to the input jacks)
   C. Outputs of luminance/chroma circuits
   D. SP record amp output, LP/SLP record amp outputs
   E. Video heads
7. Record and play
   A. Video heads
   B. Common luminance and chroma circuits to record and play
   C. RF converter

The basics of servos

We are grateful to Sencore for supplying the following information on VCR servo systems (Fig. 6-19). VCR servos control the movement of both the video tape and the video heads. Servos are a combination of mechanical devices and electronic circuits. The capstan servo controls the movement of the video tape through the VCR, while the drum servo controls the movement of the video heads.

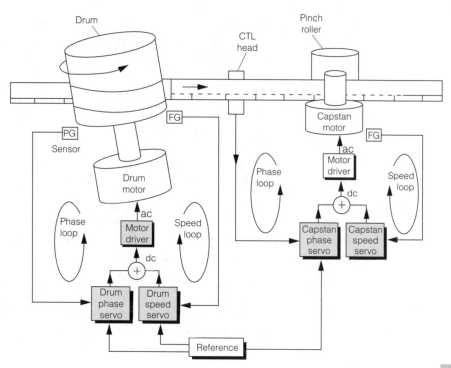

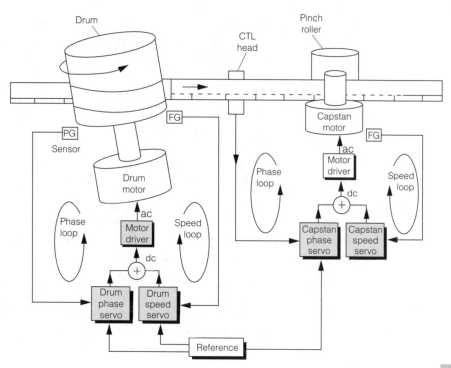

■ **6-19** *Overall block diagram.* *(Sencore)*

Together, they ensure that the correct video head is positioned exactly over the corresponding video track on the magnetic tape. Because the video heads spin close to 1800 RPM, and the recorded track is only a few thousandths of an inch wide, servo operation must be exact.

The capstan servo uses a motor and pinch roller to pull the video tape through the VCR. An electric motor driver supplies the current to run the motor. In order for the motor to run at the correct speed, a servo control "loop" monitors the motor rotation and another loop monitors the tape position. They supply correction signals to the motor driver to correctly position the video heads to pick up a signal.

A frequency generator (FG) sensor, located next to the motor, develops a signal for the servo speed control loop to tell it how fast the motor is turning. The speed servo loop compares this signal to a reference signal and sends a correction voltage to the motor driver to correct for any motor speed variations.

A second signal, called the *control track logic (CTL)* signal, is obtained from the video tape using a CTL head. It tells the capstan phase loop where the tape is at any instant in time. The CTL signal is compared to the reference signal and a correction signal is sent to the motor driver to speed up or slow down the motor to get the tape in the correct position at the correct time.

The drum servo controls the speed of the drum motor, which rotates the video heads at a rate of approximately 1800 RPM. A similar electronic motor driver supplies the current to run the drum motor. In order for the drum motor to run at the correct speed, two control "loops" are again used to monitor the speed of the spinning heads and their position.

Like the capstan servo, the drum servo uses an FG sensor to create a signal that indicates how fast the drum is turning. This signal is monitored by the drum speed loop and a correction voltage is created to speed up or slow down the motor.

The drum servo uses a pulse generator (PG) signal to tell the drum servo phase loop where the video heads are in their rotation. The PG signal is compared to the same reference signal used by the capstan circuit to create a correction voltage that places the video heads at the correct position at any instant of time.

Some TV terms from the experts

We are grateful to Hewlett-Packard for supplying the following information on technical terms. The following glossary of TV terms is taken from a Hewlett-Packard catalog. How many of these could you have identified?

How about "bruch sequencer" and "gen-lock?" Anyway, Hewlett-Packard defined these in a section of their TV products and we thought it might be a good idea to review them.

APL (Average Picture Level) The average luminance level of the unblanked portion of a television line.

Bruch sequencer An arrangement of color burst signals that ensures that the starting polarity of the burst signal is the same at the start of each field for improved stability of color synchronization.

CATV Community Antenna Television.

Chrominance This term is used to indicate both hue and saturation of a color.

Chrominance signal In television, the sidebands of the modulated chrominance subcarrier, which are added to the monochrome signal to convey color information. The chrominance signal components transmit the qualities of hue and saturation, but do not include luminance or brightness.

Color bar A test signal typically containing six basic colors: yellow, cyan, green, magenta, red, and blue. This is used to check chrominance functions of color TV systems.

Color burst Refers to a burst of 8 to 10 cycles of subcarrier on the back porch of the composite video signal. This serves as a color synchronizing signal to establish a frequency and phase reference for the chrominance signal.

Color subcarrier The carrier signal whose modulation sidebands ar added to the monochrome signals to convey color information.

Composite video signal The complete video signal. For monochrome, it consists of the picture signal and the blanking and synchronizing signals. For color, additional color synchronizing signals and color picture information are added.

Differential gain The amplitude change, usually of the color subcarrier. This is introduced by the overall circuit measured in dB or percent, as the picture signal on which it rides is varied from blanking to white level.

Differential phase The phase change of the color subcarrier introduced by the overall circuit. This is measured in degrees, as the picture signal on which it rides is varied from blanking to white level.

Gen-lock A system of regenerating subcarrier and sync from a composite video source.

H rate The time for scanning one complete line, including trace and retrace. NTSC equals 1/15734 seconds (color) or 63.5 microseconds.

Hue The attribute of color perception that determines whether an object is red, yellow, green, blue, purple, or the like. White, black, and gray are not considered as being hues.

IRE unit 7.14 mV

Luminance This indicates the amount of light intensity that is perceived by the eye as brightness.

NTSC National Television Standards Committee.

PAL (Phase Alternation Line) A system in which the subcarrier derived from the burst signal is switched 180 in phase from one liner to the next. This system helps to minimize hue errors that can occur in a color transmission.

Saturation The degree to which a color is pure and undiluted by white light, distinguished between vivid and weak shades of the same hue. The less white light in a given color, the greater is its saturation.

Staircase A video test signal containing several steps at increasing luminance levels. The staircase signal is usually amplitude modulated by the subcarrier frequency and is useful for checking amplitude and phase linearities in video systems.

VITS [Vertical Interval Test Signal (NTSC) or Vertical Insertion Test Signal (PAL)] A signal that can be included during the vertical blanking interval to permit on-the-air testing of video circuitry functions and adjustments.

Practice questions

1. The sound carrier is above the video carrier when the television signal is transmitted. After the signal passes through the mixer, the:
   A. video carrier is still below the sound carrier.
   B. video carrier is above the sound carrier.

2. To replace a deflection yoke, a first consideration is the:
   A. deflection angle.
   B. weight.
   C. size of wire used.
   D. type of core material.

3. Which of the following will decrease the brightness of the trace on a CRT?
   A. The cathode is made more positive.
   B. The cathode is made more negative.

4. An ion trap is not needed with:
   A. aluminized picture tubes.
   B. a tube that has electrostatic deflection.
   C. a tube that has electromagnetic deflection.
   D. a tube that has electromagnetic focusing.

5. The white rectangle obtained when the electron beam moves left and right, and up and down is called the:
   A. pincushion.
   B. getter.

C. white square.

D. raster.

6. In order to get a sawtooth current flowing through an electro-magnetic deflection coil, the voltage across the coil will be a:

A. parabolic waveform.

B. triangular waveform.

C. sawtooth waveform.

D. trapezoidal waveform.

7. A horizontal white line across the screen can be caused by:

A. a defective vertical oscillator.

B. a shorted sync separator.

C. a loss of high voltage.

D. an open integrator.

8. The burst amplifier is keyed into conduction during:

A. monochrome-signal reception only.

B. the time the back porch of the horizontal blanking pedestal is present.

C. the time when the VITS is present.

D. the time when equalizing pulses are present.

9. The keying pulse goes to the keyed AGC circuit from the:

A. vertical output stage.

B. sync separator.

C. flyback transformer.

D. detector.

10. The boost voltage measures below normal in a certain tube-type receiver, but it rises to normal when the yoke is discon-nected. Which is correct?

A. This is an indication that the oscillator frequency is too low.

B. This is an indication that there is a heater-to-cathode short in the horizontal output tube.

C. This is an indication that the low-voltage supply has a de-fective filter.

D. This is an indication that the yoke is defective.

11. If the horizontal retrace is not blanked, the burst might show on the screen as:

A. a color strip on the right side of the screen.

B. colored speckles of snow.

12. A blocking oscillator is most nearly similar to:

A. a Clapp oscillator.

B. a Colpitts oscillator.

C. an Armstrong oscillator.

D. a multivibrator.

13. Which of the following is least likely to be used as a sweep oscillator in a television receiver?

A. Neon sawtooth oscillator

B. Transistor blocking oscillator

C. Transistor multivibrator

14. Which is correct?

A. There are two frames per field.

B. There are two fields per frame.

15. A keystone raster might be caused by a:

A. defective yoke.

B. shorted burst amplifier input circuit.

C. 120-Hz ripple in the horizontal sweep.

D. 60-Hz ripple in the horizontal sweep.

16. Saturation is determined by the:

A. amplitude of the luminance signal.

B. phase of the I and Q signals.

C. amplitude of the color signal.

D. frequency of the burst signal.

17. The automatic degaussing circuit of a color receiver uses a VDR and:

A. a varactor.

B. a thermistor.

C. an ABL.

D. an LDR.

18. Any color can be obtained with the proper amounts of hue, saturation, and:

A. phase.

B. brightness.

19. Which of the following statements is true regarding the I and Q signals?

A. The I and Q signals have the same frequency, and they are 180° out of phase.

B. I and Q signals have the same frequency, and they are 90° out of phase.

C. The frequency of the I signal is higher than the frequency of the Q signal.

D. The frequency of the I signal is lower than the frequency of the Q signal.

20. Two signals are in quadrature if they are:

    A. equal in frequency.
    B. 90° out of phase.
    C. 180° out of phase.
    D. in phase.

21. The composite color input signal to the chroma bandpass amplifier is most likely to come from:

    A. the mixer.
    B. the receiver synchronizing section.
    C. a receiver IF amplifier.
    D. a video amplifier.

22. This keying pulse to the chroma bandpass amplifier:

    A. increases the average dc level of the signal.
    B. removes the sound IF signal.
    C. removes the color burst.
    D. removes the luminance signal.

23. The chrominance signal is separated from the composite color signal and is fed to the demodulators in the:

    A. bandpass amplifier.
    B. color killer.
    C. luminance amplifier.
    D. burst amplifier.

179

24. The composite color signal does not include the:

    A. chrominance and luminance signals.
    B. AFT correction signal.
    C. synchronizing and blanking signals.
    D. burst signal.

25. A possible cause of color, but no picture is:

    A. a defective RF amplifier.
    B. a malfunctioning color-killer circuit.
    C. a misadjusted AGC control.
    D. open delay line.

26. If the beam-landing controls are not properly adjusted, it will affect:

    A. only the color picture.
    B. both the monochrome and the color picture.
    C. only the monochrome picture.

27. The circuit that automatically disables the chrominance channel when no color signal is being received is the:

    A. luminance circuit.
    B. ACC circuit.

C. ABL circuit.

D. color-killer circuit.

28. The sound takeoff point in an intercarrier receiver is most likely to be at the:

A. RF amplifier.

B. first video amplifier.

C. IF amplifier stage.

D. mixer stage.

29. In the IF stages of a television receiver, traps are used to:

A. increase the gain of the IF amplifier.

B. block sound signals.

C. adjust the shape of the IF response curve.

D. block RF signals.

30. What is HDTV?

A. Method of automatically tuning a TV receiver

B. Brand name

C. Type of TV with very high definition

D. Type of receiver with a built-in recorder

31. In a line-operated television receiver, a zener diode is used in the B+ power supply for:

A. filtering.

B. voltage regulation.

C. rectifying.

D. voltage dividing.

32. In a television receiver, the local-oscillator frequency can be varied by changing the:

A. ABL adjustment.

B. ACC adjustment.

C. fine-tuning control.

D. AGC adjustment.

33. The maximum possible gain of the video IF section is set by the:

A. trap adjustments.

B. ABL adjustment.

C. ACC adjustment.

D. AGC adjustment.

34. The phase of the color subcarrier signal is varied by turning the:

A. AGC control.

B. optimizer control.

C. tint control.

D. color control.

35. The AFT correction voltage alters the:

   A. chroma-amplifier bandpass.

   B. maximum CRT beam current.

   C. luminance delay time.

   D. local-oscillator frequency.

36. Which of the following describes the hue of a color?

   A. Wavelength of the color

   B. Amount of white mixed with the color

   C. Brightness of the color

37. Which of the following is used for the NTSC reference white?

   A. Illuminant B

   B. Illuminant C

   C. Illuminant A

38. The subcarrier frequency for color is approximately:

   A. 6 MHz.

   B. 4.25 MHz.

   C. 4.5 MHz.

   D. 3.58 MHz.

39. The frame frequency is approximately:

   A. 30 frames per second.

   B. 60 frames per second.

   C. 120 frames per second.

   D. 525 frames per second.

40. The color burst is used to:

   A. present an amplitude-modulated color signal to the receiver for demodulation.

   B. supply a rainbow of colors for use in testing the receiver.

   C. supply bursts of color to the picture tube.

   D. synchronize the color oscillator in the receiver with the color subcarrier generator at the transmitter.

41. The scanning line frequency is approximately:

   A. 60 lines per second.

   B. 30 lines per second.

   C. 15,750 lines per second.

   D. 525 lines per second.

42. The field frequency is approximately:

   A. 15,750 fields per second.

   B. 60 fields per second.

C. 525 fields per second.

D. 30 fields per second.

43. An advantage of transmitting I and Q signals instead of transmitting R-Y and B-Y signals is:

A. more realistic blue sky.

B. lower cost.

C. more realistic green grass.

D. more realistic flesh tones.

44. Which of the following signals is transmitted with two complete sidebands?

A. I signal

B. Q signal

45. Which of the following will produce a change in hue?

A. Change in the amplitude of the chrominance signal

B. Change in the phase of the chrominance signal

C. Change in the burst frequency

46. The sweep frequency of your oscilloscope is 7.875 kHz and you are observing the signal at the output of the first video amplifier. You should see:

A. two frames, including two vertical blanking pedestals.

B. two lines including two horizontal blanking pedestals.

C. two fields, but no blanking pedestals.

47. The color burst is a minimum of eight cycles located:

A. at the top of the horizontal sync pulse.

B. on the back porch of the horizontal blanking pedestal.

C. in the 25-kHz guard band at the end of the channel.

D. on the front porch of the vertical blanking pedestal.

48. Which of the following statements is not true?

A. The vertical sync pulse has a higher amplitude than the horizontal sync pulse.

B. The vertical blanking period is longer than the horizontal blanking period.

C. The horizontal lines are still being scanned during vertical retrace.

D. Only about 480 lines out of a 525-line total contain video modulation.

49. The type of transmission used with television video signals is:

A. pulse-position modulation.

B. single-sideband, suppressed-carrier transmission.

C. amplitude modulation with one complete sideband and one vestigial sideband.

D. amplitude modulation with two sidebands.

50. In the transmitted television signal:

A. the video carrier is at a higher frequency than the center frequency of the sound signal.

B. the video carrier is at a lower frequency than the center frequency of the sound carrier.

51. The number of active scanned lines for a frame is:

A. 125 to 250.

B. 250 to 260.

C. 470 to 480.

D. 510 to 515.

52. The video carrier of a composite signal is:

A. 0.25 MHz below the upper end of the channel.

B. 0.25 MHz above the lower end of the channel.

C. 1.25 MHz below the upper end of the channel.

D. 1.25 MHz above the lower end of the channel.

53. The maximum deviation of the sound signal for television is:

A. 10 kHz.

B. 25 kHz a total carrier swing of 50 kHz.

C. 75 kHz a total carrier swing of 150 kHz.

D. 80.8 kHz.

54. The center frequencies of the FM sound signal and the video carrier frequency are:

A. 4.5 MHz apart.

B. 4.5 kHz apart.

C. 45 MHz apart.

D. 45 kHz apart.

55. The ideal response curve of a television receiver causes the video signal at the carrier to be reduced in amplitude by 50 percent. This is required in order to:

A. prevent overemphasis of lower video frequencies because they are transmitted as a vestigial sideband.

B. reduce crosstalk between the video carrier and the sound carrier of the channel being received.

C. reduce crosstalk between the video carrier and the sound carrier of the adjacent channel.

56. Select the correct statement:

    A. Audio is transmitted single sideband in the television signal.

    B. The field frequency is 1/500 of the horizontal frequency.

    C. The horizontal sweep frequency is 500 times the vertical sweep frequency.

    D. The field frequency and the vertical sweep frequency are the same value.

57. The selectivity and sensitivity of a tuner is governed primarily by the:

    A. RF amplifier stage.

    B. method of selecting tuned circuits.

    C. mixer stage.

    D. method of coupling between the RF amplifier and the mixer.

58. Which of the following types of detectors would not be used in the AFT circuit of a fully transistorized receiver?

    A. Discriminator

    B. Ratio detector

    C. Gated-beam detector

59. A limiter circuit is associated with:

    A. discriminators.

    B. gated-beam detectors.

    C. de-emphasis circuits.

    D. ratio detectors.

60. The signal for vertical-retrace blanking goes to the:

    A. video IF amplifier.

    B. video detector.

    C. video amplifier.

    D. AGC circuit.

61. The top of the picture is stretched out. You should first:

    A. replace the yoke.

    B. adjust the vertical linearity control.

62. A certain receiver will not hold sync horizontally or vertically. Other than this, the picture looks normal. A likely cause is a:

    A. defective yoke.

    B. defective sync separator.

63. Flyback pulses are used to isolate:

    A. audio.

    B. the burst signal.

C. vertical sync.

D. detector time on.

64. A dim picture tube can mean:

A. low-amplitude sync pulses.

B. loss of vertical sync pulses.

C. excessive current drawn from a pulse forming network.

D. a gassy picture tube.

65. Noise in the RF amplifier will cause:

A. snow in the picture.

B. excessive video drive.

C. pairing of the interlace.

D. None of these choices is correct.

Answers to practice questions

Question	Answer
1.	B
2.	A
3.	A
4.	A
5.	D
6.	D
7.	A
8.	B
9.	C
10.	D
11.	A
12.	C
13.	A
14.	B
15.	A
16.	C
17.	B
18.	B
19.	B
20.	B
21.	D
22.	C
23.	A
24.	B
25.	D
26.	B

Question	Answer
27.	D
28.	B
29.	C
30.	C
31.	B
32.	C
33.	D
34.	C
35.	D
36.	A
37.	B
38.	D
39.	A
40.	D
41.	C
42.	B
43.	D
44.	B
45.	B
46.	B
47.	B
48.	A
49.	C
50.	B
51.	C
52.	D
53.	B
54.	A
55.	A
56.	D
57.	A
58.	C
59.	A
60.	C
61.	B
62.	B
63.	B
64.	D
65.	A

Test equipment
and troubleshooting

7

Associate level *Important review material for Sections VI, VII, and VIII of the Associate-level CET test.*

Journeyman level consumer option *Important review material for Sections XIV and XV of the Journeyman-level CET test.*

Section 15 of the Consumer Electronics test is called *Test Equipment*. It contains 5 questions. Section 14 is called *Troubleshooting Consumer Equipment*, and it contains 15 questions. This chapter reviews the types of questions you will find in those two sections. The practice tests in the Appendices give a more thorough review of the types of questions about instruments and troubleshooting.

Traditionally, Journeyman technicians have done very well answering questions about these subjects. In the first place, anyone taking these sections is presumed to have four years of equivalent practical experience. So, the person taking the Journeyman test is very likely to be familiar with the test equipment and test procedures used for television servicing.

Associate-level technicians should also do well on this material. It is well covered in textbooks and in the first two years of post-secondary school courses.

It is important for you to understand the distinction made in the CET test between tests and measurements, and troubleshooting. *Tests and measurements* includes the setups, adjustments, tests, and measurements that you might make in preparation for troubleshooting.

For example, you might be asked how to obtain the response curve for an amplifier. That is actually a test setup. Or, you might be asked a question on how to measure the rise time of a pulse with an oscilloscope. That would be a measurement. You will get a good idea of the types of questions in each of the practice test sections in the Appendices.

To answer questions in the troubleshooting section, a technician must be able to recognize symptoms, and must be able to perform a logical troubleshooting procedure to locate a defective component or part.

In the actual test, you might be shown the schematic of a system and asked the cause of incorrect voltage or waveform measurements at some point. Those questions are no different from questions about individual circuit problems.

In some cases, you might be asked about making repairs. For example, you might be asked about the type of solder used to replace surface-mount devices, the use of heatsinks, or the use of substitute replacement parts.

Expect questions on troubleshooting logic circuits using a logic probe, and about the use of oscilloscope and voltmeter probes. Be sure you understand the use of timing diagrams for logic circuits.

Meter movements

A galvanometer is a very sensitive current-measuring device. It is not uncommon to have galvanometers with a full-scale current of 10 or 50 microamperes. The deflection sensitivity of a meter is the reciprocal of its full-scale measuring ability. That sensitivity is stated in ohms per volt. Mathematically:

$$Ohms\ per\ volt = \frac{1}{Full\text{-}scale\ current}$$

This is a very important point: the ohms-per-volt rating of a meter has nothing to do with the number of ohms impedance that the meter offers to the circuit when making a particular voltage measurement! The impedance offered by a meter when making a voltage measurement is not found by multiplying the measured voltage by the ohms-per-volt rating.

The calculations of meter series multiplier resistance and meter shunt resistance is important for understanding the use of meters. However, calculations for those values are not usually required on a CET test. Remember that a *multiplier* is used to convert a sensitive analog meter movement into a voltmeter. A *shunt* is used to convert a sensitive analog meter movement into an ammeter or milliammeter.

Practice question

A certain meter movement is rated at 100,000 ohms per volt. This means that:

A. the impedance of the meter is 100,000 ohms for each volt being measured.
B. the full-scale deflection of the meter movement is 10 microamperes.

Answer: Choice B is correct. Remember that current equals volts divided by ohms. Here is how ohms-per-volt is converted to current:

$$\frac{1}{\text{ohms per volt}} = \frac{1}{\text{ohms/volt}} = \frac{\text{volts}}{\text{ohms}} = \text{amps}$$

The effect of a meter on a circuit

When you make a measurement with a voltmeter the resistance of the meter changes the conditions of the circuit. The meter will display less voltage than the true voltage in the circuit.

Consider the circuit in Fig. 7-1. The values are chosen to demonstrate the effect of the measurement as demonstrated in the following practice questions.

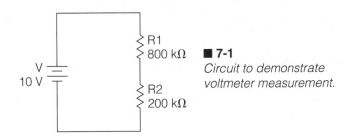

7-1
Circuit to demonstrate voltmeter measurement.

Practice question

What is the true voltage across R_2 the circuit of Fig. 7-1?

Answer: The true voltage can be determined by the proportional method:

$$V_2 = V\left(\frac{R_2}{R_1 + R_2}\right) = 2 \text{ V}$$

Practice problem

You have a meter movement rated at 20,000 ohms per volt. The resistance of the meter movement is 27 ohms. You want to use it to make a voltmeter that can measure in the range 0–10 V. What is the resistance of the series multiplier (R_X) resistor that must be used with the meter movement?

Answer: The full-scale (FS) current rating of the meter movement is:

$$FS\ current = \frac{1}{20,000} = 50\ microamperes$$

That is the amount of current that will flow through the meter movement when it is measuring 10 V. See Fig. 7-2. By Ohm's law:

$$R_M + 27 = \frac{V}{I_M}$$

Therefore, the meter multiplier resistance is +199,973 ohms.

To obtain that resistance it would be necessary to use laser trimming. Important: Observe that the total resistance offered by the meter branch is:

$$199,973 + 27 = 200,000\ ohms!$$

Practice question

Using the meter you designed for Fig. 7-2, determine what voltage that meter will display when measuring the voltage across R2 in the circuit of Fig. 7-1.

Answer: Figure 7-3 shows the equivalent circuit with the meter across R2. What you are really looking for is the voltage across R2. Again, using the proportional method of finding the voltage across the parallel resistance(R_P) of 100 k ohms.

To summarize, the meter displays only 1.11+ volt, but the actual voltage (without the meter) is two volts.

If you are not adverse to short-cut mathematics, you can calculate the resistance of the voltmeter by using the ohms-per-volt rating of the meter movement and the full scale voltage.

   Meter resistance = Full-scale voltage ÷ Full-scale current

For this example:

$$Resistance\ of\ meter = \frac{10\ V}{0.000050}$$

$$V_2 = 10\left(\frac{100}{100 + 800}\right) = 1.11 + V$$

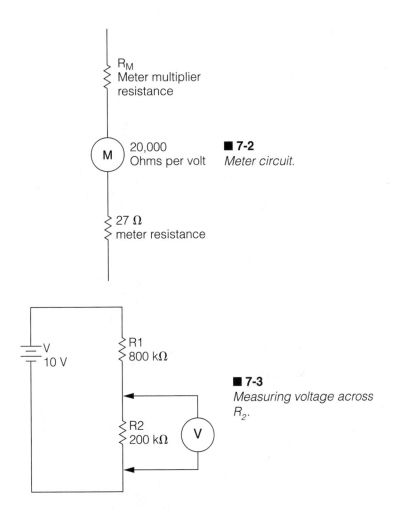

R_M
Meter multiplier
resistance

(M) 20,000
Ohms per volt

27 Ω
meter resistance

■ 7-2
Meter circuit.

V
10 V

R1
800 kΩ

R2
200 kΩ

(V)

■ 7-3
*Measuring voltage across
R_2.*

Analog meter movements can be divided into two basic groups—
those with jeweled bearings and those that use a taut band. Figure
7-4 compares these two movements.

Figure 7-4A shows the side view of a jeweled bearing. Notice that
this bearing fits inside of a cup. As the meter moves upscale, this
bearing supports the pointer as it turns in the cup. Hopefully, it
will turn with a minimum amount of friction. When the meter is
new, there is very little problem here. However, as the meter gets
older, certain disadvantages of this type of movement become ap-
parent.

One disadvantage is that the bearing eventually begins to wear
and no longer fits snugly with the minimum contact area in the
cup. When that happens, the meter begins to stick at various
points. For that reason, technicians are advised to tap the meter

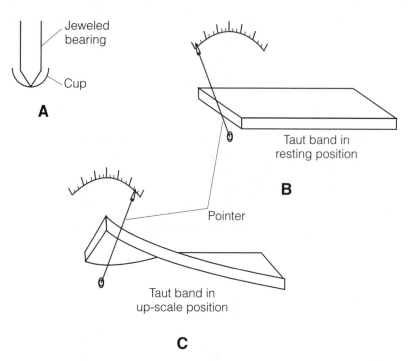

Jeweled bearing

Cup

A

Taut band in resting position

B

Pointer

Taut band in up-scale position

C

■ **7-4** *Taut band at rest (B), and defecting upscale (C).*

(gently!) when making a measurement with a meter that uses a jeweled movement.

It should be obvious that if you stand this type of meter on end, the jewel is going to ride toward the edge of the cup rather than in the center. Again, inaccuracy is possible. Remember that the desirable minimum friction occurs when the meter is laid on its back rather than when it is standing up.

If the jeweled-type meter movement is subjected to a strong jolt (which could occur from dropping it), permanent damage might occur to this simple movement and the meter will no longer be useful. In the jeweled meter movement, a small spring is used for the restoring force, to return the pointer to zero when the current stops.

The taut band meter movement uses the twist of a band as the pointer is moved upscale. Figure 7-4B shows the taut band in its resting position and the band as the pointer moves upscale. This band restores the pointer to its resting position by moving it back to zero when the measurement current is completed. The taut band movement is considered to be much more reliable and much more rugged. It is used in the more rugged meters.

To convert a meter movement to an ohmmeter function, it is necessary to supply a battery voltage and either a series multiplier resistor or a shunt resistor to protect the meter during the measurement. Because the battery supplies a current to the circuit or component being measured, you must be careful not to overload sensitive diodes and transistor junctions when measuring their resistance. In many multimeters (also called *volt-ohm-milliammeters*), this is most likely to occur when the meter is in the $R \times 1$ position.

Digital meters are much more accurate and convenient to use than analog types. Furthermore, they do not have the problem of *parallax*, which is present in the analog type. Parallax occurs when you view the meter pointer from an angle, rather than from directly above.

To reduce the problem of parallax, some analog meters have a mirrored scale. Its purpose is to assure that the measurement is taken with the viewer directly above the meter pointer. Then, the pointer and its reflection in the mirror should overlap. That can only occur if you're directly above the pointer. From that position, you make the measurement.

Oscilloscopes

As a general rule, triggered-sweep scopes are preferred over recurrent-sweep scopes for advanced electronic servicing. One advantage of the triggered-sweep scope is that the sweep time can be calibrated very accurately in microseconds (or milliseconds) per centimeter (or inch). This makes it very easy to find the time (T) for one cycle of an input waveform. By using the equation:

$$f = \frac{1}{T}$$

it is possible to calculate the frequency (f) of the signal being displayed when the time for one cycle (T) is known.

In order to calculate the frequency of the display with the recurrent-sweep scope, it is necessary to use some type of time marker. One way is to apply accurate narrow timing marker pulses to the Z axis. Remember that the Z axis of the oscilloscope controls the brightness of the display.

Another advantage of the triggered-sweep scope is that it will remain synchronized with the displayed incoming signal—even if the frequency of that signal varies slightly. The recurrent sweep oscilloscope is not very stable against variations in frequency of the waveform being measured.

Be sure that you can calculate the frequency ratios of two sine waves in a Lissajous pattern. The simplest way to do this is illustrated in Fig. 7-5. The method used to obtain a Lissajous pattern on an oscilloscope is also shown. The number of times that the display touches the horizontal line is directly related to the frequency of the vertical signal (or the frequency on the vertical terminals).

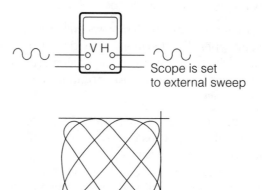

Scope is set
to external sweep

■ **7-5** *Oscilloscope hookup for Lissajous display.*

The number of times that the display touches the vertical line (V) is related to the frequency of the signal being applied to the horizontal input (H) terminals. Remembering this simple fact from Fig. 7-5, you can write the equation:

$$\frac{V}{H} = \frac{4}{3}$$

The result shows that the vertical frequency is 4/3 times the horizontal frequency. So, if the horizontal frequency is 300 Hz, the vertical frequency will be:

$$Vertical\,frequency = \frac{3}{2} \times 300 = 450 \text{ hertz}$$

A Lissajous pattern can also be used to show the phase difference between two sine-wave signals. The test setup is the same as the one shown in Fig. 7-5. The technique is based on the measurements shown in Fig. 7-6. Using these measurements, the equation for the phase angle is:

$$\text{arc sin } \phi = \frac{A}{B}$$

The term *arc sin* is also written as "sin⁻¹." Do not interpret this symbol as meaning sin raised to the –1 power! The term *arc sin* (or, sin⁻¹) simply means the angle that has a sine equal to that ratio. For example, arc sine 0.5 = 30°. The arc sine on calculators is usually called *Inv sin*.

Remember that a Lissajous pattern that compares two sine waves with frequencies that are equal will produce a perfect circle if the two sine waves are 90° out of phase. If they are 0° out of phase, a straight line will rise from left to right. If they are 180° out of phase, a straight line will decrease from left to right (Fig. 7-6).

$$\text{Phase angle } \phi = \text{arc sin } \frac{A}{B} = \sin^{-1} \frac{A}{B}$$

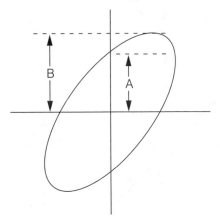

■ **7-6** *Oscilloscope display for phase measurements.*

There have been cases where the horizontal deflection plates were wired backwards. However, any question on a CET test regarding Lissajous patterns assumes that the scope is properly wired.

The use of an oscilloscope to measure frequency and phase is primarily for applications where the available test equipment is limited, or where a quick answer without high accuracy can be used. A digital frequency meter gives a much more convenient and accurate measurement of frequency. For the measurement of phase, a Z-angle meter gives a more convenient measurement with a direct digital readout.

There is always a tradeoff between the accuracy needed and the cost of more accurate measuring equipment.

Practice question

A Lissajous pattern is obtained by comparing the input sine- wave signal to output sine-wave signal of a class-A amplifier. This should produce a straight line as shown in Fig. 7-7A. However, for one particular amplifier the line is not straight (Fig. 7-7B). What is the indication?

A. The output frequency does not equal the input frequency.
B. The signals are not exactly 180° out of phase.
C. The amplifier has some form of linear distortion.
D. None of these choices is correct.

Answer: Choice C is correct. If there is a line, but it is not straight, the indication is that some form of nonlinear distortion is occurring in the amplifier.

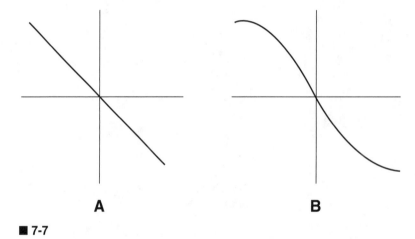

A   **B**

■ 7-7

Remember that you can use an oscilloscope to measure current instead of voltage. The technique is to place a 1- or 10-ohm resistor in series with the current being measured. The scope is then used to measure the voltage across that resistor. By using a 1-ohm resistor, the current displayed on the voltmeter (or, oscilloscope) is equal to the voltage. If you use a 10-ohm resistor, the display readout in volts must be divided by 10 to get the numerical value of current.

Triggered-sweep scopes are generally designed with an internal calibration for voltage measurements. To use a recurrent-sweep scope for measuring voltage, it is usually necessary to use some form of the voltage calibrator to help make the measurement.

Using probes

It is sometimes necessary to use special probes to make measurements with an oscilloscope or voltmeter. Many inexpensive oscilloscopes cannot display higher frequencies, such as the intermediate frequency of a basic AM receiver, let alone the IF of a television. To get around this problem, a detector probe is normally used. It shows the envelope of the IF signal, but not the IF signals that make up that envelope. Presumably, if the envelope is not correct, the IF stage is not operating properly.

High-voltage probes are used with voltmeters. They are generally marked with the amount of multiplication necessary to convert the voltmeter reading to the actual value being made. For example, if a ×10 (times 10) probe is used to make a measurement, and the voltmeter indicates 100 volts, the actual value being measured is $100 \times 10 = 1000$ volts.

Square-wave test

Be sure that you are familiar with the square-wave test of an amplifier. In this test, a square wave is applied to the input of the amplifier and the output is observed on an oscilloscope. This test should never be used with an amplifier that has an inductive load. There are two reasons for this. One is that the inductive kickback could destroy an output transistor unless the designer has made some special provision against this possibility. The second reason is that the reactive load will change the shape of the square wave and might give an erroneous indication of poor performance.

The most important output square-wave patterns to know are illustrated in Fig. 7-8. Observe that an overshoot on the square waveform is an indication of excessive high-frequency response, but, it also might indicate an inductive load.

Ringing test

This characteristic is utilized in the ringing test of test equipment that is designed to evaluate TV picture tube yokes, transformers, and other inductive components. With a ringing test the inductor is electrically shock-excited with a pulse. Assuming that the inductor is not short circuited, it should go into self oscillation with its disturbed capacitance. The result is a high number of cycles of ringing. However, if the inductor is defective only one or two (or maybe three) cycles of ringing will be observed.

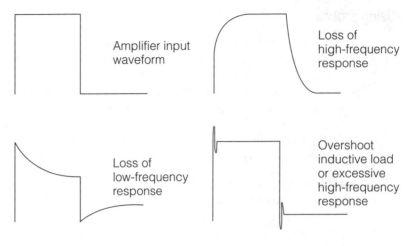

Amplifier input waveform

Loss of high-frequency response

Loss of low-frequency response

Overshoot inductive load or excessive high-frequency response

■ **7-8** *Results of square-wave test.*

The ringing test is a quick way to evaluate the ability of the inductor (like a deflection coil) to do its job. If you want the inductance value it should be measured on a bridge or (indirectly) on a Q-meter or an LCR meter.

Evaluating parameters

As a technician, you should be very familiar with the methods used to evaluate bipolar transistors and FETs. The various parameters are defined in Table 7-1. Vacuum tubes are included in this table to emphasize the similarities between tubes and FETs. Remember, though, there will be no questions on tubes in the CET tests.

Pay attention to the difference between the ac and dc parameters used to measure transistor performance.

Alpha (h_{FB}) is a parameter obtained with the transistor in the common-base configuration, but beta (h_{FE}) is a parameter that is obtained with the transistor in the common-emitter configuration.

A quick test for transistor operation is to use an ohmmeter with the procedure described in Fig. 7-9. Be careful not to use the ohmmeter on the R × 1 scale of a volt-ohm-milliammeter (VOM) because, with some meters, that can readily burn out a diode or transistor junction.

■ Table 7-1 Bipolar transistor, FET, and vacuum-tube parameters.

As a general rule, the dc characteristics are noted with capital letters and the ac (or "dynamic") characteristics are noted with lowercase letters. The symbol Δ (delta) means "a small change in value." Thus, ΔV_g means a small change in grid voltage.

	Name	Symbol	Description	Equation
Tubes	Amplification factor	μ	Measure the ability of the tube to amplify	$\mu = \dfrac{\Delta V_p}{\Delta V_g} \Big] I_o = \text{Constant}$
	Plate resistance	r_p	Plate-to-cathode resistance of the tube	$r_p = \dfrac{\Delta V_p}{\Delta I_p} \Big] V_g = \text{Constant}$
	Transconductance (Also known as mutual conductance)	g_m	Change in plate current produced by a change in grid voltage	$g_m = \dfrac{\Delta I_p}{\Delta V_g} \Big] V_p = \text{Constant}$
	Relationship between μ g_m and r_p			$\mu = g_m\, r_p$
Bipolar transistors	dc Alpha	α_{dc}	Common-base forward current transfer ratio	$\alpha_{dc} = \dfrac{I_C}{I_E}$
		h_{FB}	Same as dc alpha	$h_{FB} = \alpha$
	dc Beta	β_{dc}	Common-emitter forward-current transfer ratio	$\beta_{dc} = \dfrac{I_C}{I_B}$
		h_{FE}	Same as dc beta	$h_{FE} = \beta$
	Gain-bandwidth product	fr	Frequency at which $\beta = 1$	

■ Table 7-1 Continued.

Name	Symbol	Description	Equation
ac Alpha	α_{ac}	Common-base forward-current transfer ratio using signals or current changes (instead of dc currents as in αDC)	$\alpha = h_{FB} = \dfrac{\Delta I_C}{\Delta I_E} = \dfrac{i_c}{i_E}$
ac Beta	β_{ac}	Common-base forward-current transfer ratio using signals or current changes (instead of dc currents as in βDC)	$\beta = h_{FE} = \dfrac{\Delta I_C}{\Delta I_B} = \dfrac{i_c}{i_b}$
Relationship between α and β			$\alpha = \dfrac{\beta}{1+\beta} \quad \beta = \dfrac{\alpha}{1-\alpha}$
Transconductance	g_m	Similar to the transconductance of vacuum tubes	$g_m = \dfrac{\Delta I_D}{\Delta V_{GS}} \, V_{DS} = \text{Constant}$
FETs			

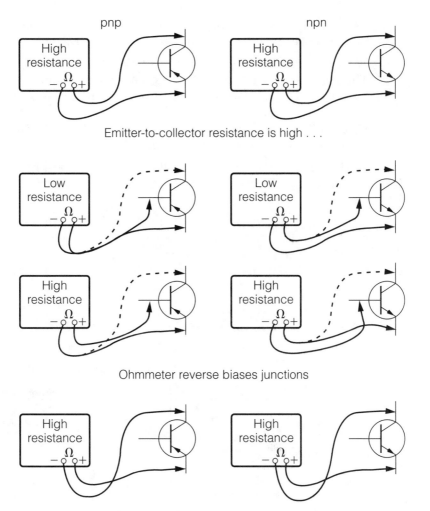

pnp   npn

Emitter-to-collector resistance is high . . .

Ohmmeter reverse biases junctions

■ **7-9** *Quick test for transistor operation.*

Testing amplifiers

A quick test that is sometimes used to determine if a bipolar transistor amplifier is operating is illustrated in Fig. 7-10A. In this test, the emitter-to-base junction is shorted while the collector voltage is observed. Theoretically, at least, shorting the emitter to the base of the transistor does not hurt it. It simply removes the forward bias; therefore, it prevents the transistor from conducting. A voltmeter measurement is made to ensure that the transistor has been cut off.

If the transistor shuts off with the emitter-base junction shorted, it means that there has been control over the collector current by

Test equipment and troubleshooting

the transistor base. Obviously, if the collector current stops flowing, the collector voltage should go to the power supply voltage (30 V, in this case) for the setup in Fig. 7-10A.

This test is somewhat controversial because in the hands of an inexperienced technician, some erroneous data can be obtained. For example, on transistors that are direct coupled, as shown in Fig. 7-10B, the bias of the second transistor (Q2) is obtained directly from the collector voltage of Q1.

If Q1 is cut off, the base of Q2 will go to a highly positive value. That, in turn, will destroy the second transistor in that direct-coupled configuration.

Another problem with the short-circuit quick test of a transistor amplifier is that some transistors are operated in the class-B configuration as in Fig. 7-10C. Remember that in a class-B configuration, the bias between the emitter and the base is 0 volts (or nearly

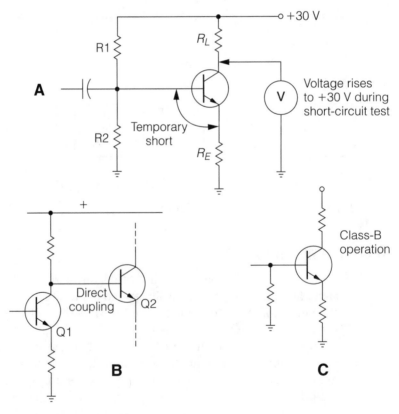

■ **7-10** *Circuits to explain transistor tests.*

so). Shorting the emitter to the base in that configuration might not produce any observable change in the collector voltage. As a matter of fact, the collector voltage will already be at its maximum positive value in the absence of an input signal.

In order to properly evaluate a collector voltage measurement, you must understand the transistor configuration. Specifically, the circuit must be viewed in relation to the power-supply voltage.

Transistor configurations

Figure 7-11 shows two ways to connect a transistor across a dc power supply, with regard to power-supply polarity. Both are class-A amplifiers. Observe the differences in voltmeter readings when you apply the short-circuit test in each circuit. This is another criticism of the short-circuit test used for transistor amplifiers.

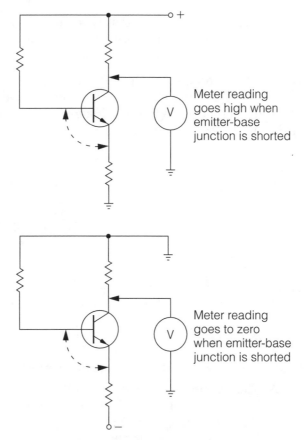

■ **7-11** *Readings with different circuit hookup.*

However, despite the various disadvantages that have been covered so far, the short-circuit test is considered to be a valuable troubleshooting technique and it is used by many technicians in the field.

Some types of equipment are marked on the printed circuit board to show the technician where the short-circuit test can be safely applied. This relieves the technician of the responsibility of making sure that the proper configuration is being tested.

Crossover distortion

Transistors in push-pull and totem-pole configurations are normally operated in a class-AB configuration. The reason is that a small amount of forward bias is necessary to eliminate the crossover distortion that can occur as one transistor stops operating and other one begins.

The necessary forward bias that is obtained with class-AB operation prevents one transistor from reaching cutoff before the other transistor begins to conduct.

Crossover distortion is easily observed. Apply a pure sine wave to a class-B biased push-pull or totem-pole configuration and observe the output across the load. Figure 7-12 shows the output signal with crossover distortion.

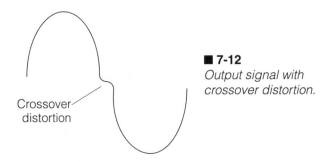

Crossover distortion

■ **7-12**
Output signal with crossover distortion.

Practice questions

1. Which of the following statements is correct?
   A. The forward-bias current for a transistor that is operating class B will be greater than the forward bias on a transistor that is operating class A.
   B. The forward-bias current for a transistor that is operating class A will be greater than the forward bias on a transistor that is operating class B.

2. Which of the following is true?

   A. Shorting the base of a bipolar transistor to the emitter will surely destroy the transistor if it is operating as a class-A amplifier.

   B. Shorting the base of a bipolar transistor to the collector in a class-A amplifier will surely destroy a transistor.

3. To check a diode with a VOM, set the ohmmeter scale to the:

   A. R times 1 ($R{\times}1$) position.

   B. R times 100 ($R{\times}100$) position.

4. Which of the following is not the correct symbol for ac beta?

   A. h_{fe}

   B. h_{FE}

5. In the normal operation of a bipolar transistor, which junction is normally forward biased?

   A. The collector-base junction

   B. The emitter-base junction

6. The reverse-current in the collector-base junction of a bipolar transistor is called:

   A. I_{CER}

   B. I_{CEO}

   C. I_{ECS}

   D. I_{CBO}

7. In an ac resistance circuit, the product of the RMS volts and RMS amperes gives the:

   A. average power.

   B. true power.

   C. cosine of the phase angle.

   D. VARS.

8. A 1.5-volt dry cell (Le Clanche cell) has been discarded because it does not provide enough energy to operate a flashlight. An FET meter with a high input resistance will measure:

   A. less than 0.5 volt across the cell terminals.

   B. about 1.5 volts across the cell terminals.

   C. zero volts across the cell terminals.

   D. 0.99 volt across the cell terminals.

9. Which of the following can be turned on with a gate signal, but cannot be turned off with a gate signal?

   A. MOSFET

   B. JFET

C. SCR

D. tunnel diode

10. You can get a negative output voltage from a half-wave rectifier power supply that has a positive output voltage by simply reversing the diode. This statement is:

A. true.

B. false.

11. Assume that the grid voltage of a television CRT is held to a constant dc voltage. The brightness of the screen will increase if the cathode dc voltage is made:

A. more positive.

B. less positive.

12. For a half-wave rectifier circuit operating from a 120-Vac power line, the approximate output voltage should be:

A. 110 V.

B. 95 V.

C. about 165 V.

D. 55 V.

13. To prevent crossover distortion in push-pull power transistors:

A. supply a low forward (emitter-to-base) bias to each transistor base.

B. supply a low reverse (emitter-to-base) bias to each transistor base.

14. When the collector current of a bipolar transistor flows through an inductor (like the coil of a relay), the transistor can be destroyed by inductive kickback. To prevent that, which of the following can be connected across the inductor?

A. VDR

B. Thermal relay

C. Hall device

D. Slow-blow fuse

15. When semiconductor diodes are connected in parallel, it is possible for one diode to conduct and prevent the other one from conducting. This is called *current hogging*. It can be prevented by:

A. connecting high-resistance resistors in series with each diode.

B. connecting low-resistance resistors in series with each diode.

16. For an enhancement MOSFET, the gate voltage is:

    A. the same polarity as the drain voltage.

    B. opposite in polarity to the drain voltage.

17. The voltage across a certain 47-kΩ resistor is measured at 7.5 V and the current through the resistor is known to be 160 microamperes. If the resistor tolerance is 1%, should the resistor be replaced?

    A. Yes

    B. No

18. Which of the following is true for a common gate FET amplifier?

    A. The output signal is in phase with the input signal.

    B. The output signal is 180° out of phase with the input signal.

19. For a two-input AND gate, the output is logic 0 when:

    A. both inputs are at logic 0.

    B. both inputs are at logic 1.

    C. one input is at logic 1 and the other input is at logic 0.

20. An audio frequency can be determined by comparing the audio input voltage and the output voltage of a signal generator using an oscilloscope. The pattern is called:

    A. Lissajous.

    B. Cisoid.

    C. Remington.

    D. None of these choices is correct.

21. Which of the following is NOT true?

    A. Wattmeters usually measure voltage and current.

    B. Voltmeters offer a high resistance to current.

    C. Milliammeters and other current meters must not be connected across a voltage.

    D. Ohmmeters can be connected across a voltage, but not in series with a current.

Answers to questions

Question	Answer
1.	B
2.	B
3.	B
4.	B
5.	B

Question	Answer
6.	D
7.	A
8.	B
9.	C
10.	B (The electrolytics must be reversed also!)
11.	A
12.	C
13.	A
14.	A
15.	B
16.	A
17.	B
18.	A
19.	A
20.	A
21.	D

Practice questions

22. To determine the reverse resistance of an NPN collector-base junction:

    A. place the positive lead of the ohmmeter on the collector terminal and the negative lead on the base terminal.

    B. place the negative lead of the ohmmeter on the collector terminal and the positive lead on the base terminal.

23. One difference between a triggered-sweep scope and a recurrent-sweep oscilloscope is that:

    A. the triggered-sweep scope does not have a trace.

    B. the triggered-sweep scope cannot be used for voltage measurements.

24. Which of the following explains the use of markers on an oscilloscope trace?

    A. They blank the scope during retrace.

    B. They identify certain amplitudes on the response curves.

    C. They identify certain frequencies on the response curve.

    D. They eliminate the back voltage generated in the yoke during retrace.

25. The ripple frequency of a full-wave rectifier circuit in a 60-Hz power system would be:

    A. 60 Hz.

    B. 120 Hz.

C. 400 Hz.
D. 800 Hz.

26. An oscilloscope is calibrated for a 10 volt-per-inch vertical deflection. If a 5-volt RMS sine-wave voltage is fed to the vertical terminals of the scope, the deflection should be:

A. exactly 1.0 inch.
B. about 2.8 inches.
C. about 1.11 inches.
D. about 1.4 inches.

27. If a pure sine wave is applied to an RC differentiating circuit, the output will be:

A. pulses.
B. a square wave.
C. a triangular waveform.
D. a sine waveform.

28. A transistor junction might be damaged by excessive current flow when an ohmmeter is used to measure the forward resistance. That is most likely to occur when the ohmmeter is in the:

A. R-x-1 position.
B. R-x-10-kΩ position.

29. When soldering parts into radios, use:

A. 60–40 solder.
B. 40–60 solder.

30. The collector of a power transistor can be connected:

A. through the smallest terminal lead.
B. through the case or through a threaded stud.

31. A true square wave would be obtained by:

A. combining a fundamental-frequency sine wave with an infinite number of odd harmonic sine waves—each having the proper amplitude and phase.
B. combining a fundamental-frequency sine wave with an infinite number of even harmonic sine waves, each having the proper amplitude and phase.

32. Which of the following is the lower value?

A. RMS value of a sine-wave voltage
B. Average value of a sine-wave voltage

33. A sine-wave voltage is measured with an oscilloscope and found to be 15 volts peak to peak. What is the RMS value of this voltage?

    A. 10.6 volts
    B. 9.54 volts
    C. 7.2 volts
    D. 5.3 volts

34. To get a Lissajous pattern on an oscilloscope:

    A. feed a sine-wave voltage to the vertical deflection plates and a sawtooth wave to the horizontal deflection plates.
    B. feed a sine-wave voltage to both the vertical and the horizontal deflection plates.
    C. feed a sine-wave voltage to the horizontal deflection plates and a sawtooth wave to the vertical deflection plates.

35. An oscilloscope will display a circle when:

    A. two sine waves that are 90° out of phase are fed to the vertical and horizontal deflection plates.
    B. two in-phase sine waves are fed to the horizontal and vertical deflection plates.

36. The current drawn by a transistor radio is measured by connecting a milliammeter across the on/off switch. Which of the following is true?

    A. The switch must be in the on position.
    B. An ohmmeter should be connected across the switch.
    C. If the radio has a class-B audio amplifier, the current drain will decrease when the radio is tuned to a station.
    D. If the radio has a class-B audio amplifier, the current drawn will increase when the radio is tuned to a station.

37. The forward emitter-base resistance on a certain transistor is 200 ohms, and the reverse resistance is 500 ohms. Which of the following is correct?

    A. The ratio is too high; the transistor is not good.
    B. The ratio is too low; the transistor is not good.

38. The base of an NPN transistor in a class-A amplifier circuit has a voltage of –6.8 volts. Which of the following would you expect to measure on the emitter?

    A. +3 volts
    B. +7.2 volts
    C. –7.2 volts
    D. –3 volts

39. An ohmmeter is connected across the emitter-collector terminals of a transistor. When the collector and base leads are shorted together, the ohmmeter shows a lower resistance. Which of the following is true?

A. The transistor is shorted.
B. This is normal for a good transistor.
C. The emitter-base junction of the transistor is open.
D. This can never happen.

40. The collector circuit of a certain PNP transistor is connected to ground through a transformer winding. You would expect the emitter of this transistor to be:

A. positive, with respect to ground.
B. negative, with respect to ground.

41. The bias on a certain transistor measures too low, but its collector current is too high. A likely cause is:

A. an open voltage divider circuit for biasing the base.
B. a leaky transistor.
C. an open bypass capacitor in the emitter stage.
D. an open emitter resistor.

42. A certain transistor is not conducting. The base bias voltage is measured and found to be OK. A possible cause is:

A. the power supply is not operating.
B. the transistor is open.
C. the transistor is shorted.
D. the power supply is at fault.

43. When a certain voltmeter rated at 20,000 ohms per volt is connected across a 10-V voltage, full-scale deflection occurs. If the meter is on the 10-V scaler, how much meter current is flowing?

A. 500 milliamperes
B. 50 microamperes
C. 5 microamperes
D. 500 microamperes

44. If you want to look at the response curve of the video IF amplifiers on an oscilloscope, use:

A. an X-Y plotter.
B. a logic probe.
C. a sweep generator.
D. a square-wave test.

Answers to test

Question	Answer
22.	A
23.	A
24.	C
25.	B
26.	D
27.	D
28.	A
29.	A
30.	B
31.	A
32.	B
33.	D
34.	B
35.	A
36.	D
37.	B
38.	C
39.	B
40.	A
41.	B
42.	B
43.	B
44.	C

Practice for taking CET tests

8

The art of test taking

Some people are not good at taking tests even though they are well prepared in the subject. Others do well even when they are not as well prepared. Because you are going to take a CET test, you might as well take a look at some of the methods that work and some that don't work.

Some say that they prefer to go through the test and answer the easiest questions first. Then, they go back and answer the most difficult questions.

That might not be such a good idea. It leaves the more difficult questions to the time when you are getting tired of taking the test. That happens. You can start out with enthusiasm, then after 45 minutes, you just get tired. That is not a good time to be answering tough questions.

Another problem is that it takes time to answer the easy questions first. It can result in your being in a hurry when you should be taking more time to analyze some of the difficult questions.

Learning to pace yourself

It is not unusual for a technician to take a 75-question test without ever trying a practice test. That is definitely not a good idea!

As you go through chapters in your study guide, you might answer questions as you go. Taking a 75-question test is harder. That is true even though the questions are the same. You need to check your ability to take a test that can take two hours. You need to be able to do that without getting so weary that you just turn your paper in to be done with it. If there is any sure way to fail a test, it is by not being able to pace yourself.

We have heard technicians claim they do better on a test if they don't check over their paper before handing it in. That is a sure sign that they are too weary to go any further. Certainly, some answers seem right at first, but it later turns out they are not the best choice.

If you have time to go over your paper before you hand it in, be very careful you don't change answers you are not sure of. In other words, if you change an answer, be sure that you know why there is a better choice!

The guessing game and the educated guess

If you are guessing more than you are analyzing, it is a sure sign you are not properly prepared in the subject. You have to be honest with yourself about this. If you don't pass and you realize you were doing a lot of guessing, do some intelligent studying before you try it again. ISCET has books and practice tests that can be very helpful. Write to them and ask for a list. The address for IS-CET is given in Chapter 1 of this book.

You are not given an extra penalty for making a wrong guess. If you are sure that you do not know the answer to a question, and you have given it some thought, make the best possible guess! Sometimes you can eliminate some obvious wrong choices and that gives you a better chance. Not only that, but if you are trained in the subject, your insight might be a help in picking a correct choice. It is known as an *educated guess*.

Give yourself an honest appraisal

Take the test in this chapter under the same conditions as you will be taking the actual CET test. Take the complete test at one sitting! Don't break for lunch! Turn off the TV set and/or the radio! Turn off the telephone!

We are giving you a 75-question Associate-level practice test in this chapter. This practice test is primarily about basic fundamentals. However, there are some questions that could also appear in a Journeyman CET test.

Any time you take a test and you find yourself falling asleep, you know you are not giving it your best try.

More about the tests you are about to take

The test in this chapter is much harder than the Associate-level test. It covers a wider range of basic theory, and it tests your ability to answer questions that explore theory at a deeper level. If you cannot answer some of these questions, it is possible that you are not really well prepared on those subjects. That means you might give a wrong answer to an easier question on the same subject.

The test is not divided into subjects like Basic Mathematics or Test Equipment. Don't let those subject headings on an actual test lull you into making a bad decision. For example, don't say something like "I was never good at math, so I'll leave this section until last." Actually, if you are an experienced technician, you will not find any one section in the actual CET test harder than another.

In the test for this chapter, the subjects are mixed together to discourage you from trying to pick out easy questions. Remember this very important thing about taking any practice test: If you can answer a question, you haven't learned anything. You have just verified that you know that subject. Of course, that is very important! However, a practice test should be a learning experience too. If you can't answer a question correctly, think of it an opportunity to broaden your knowledge. Learn something from the experience.

If you have not already taken the practice CET tests in the Appendices, save them for later. Take the tests in this chapter and the next chapter and learn to pace yourself. Some material in these tests was not covered before. Detailed answers are given for this test at the end of the chapter.

Practice test

1. Which of the following components produces a voltage that is directly related to the strength of a magnetic field?

    A. CCD
    B. Thyristor
    C. Hall Device
    D. Bead Ledge

2. Is the following statement correct? *Iron is attracted to a magnetic field, but, some materials are repelled by a magnetic field.*

    A. The statement is correct.
    B. The statement is not correct.

3. A certain type of motor turns 15° every time it receives a pulse. What is this type of motor?

    A. Synchronous motor
    B. Induction motor

C. Compound-wound motor

D. Stepping motor

4. The only difference between the two circuits in Fig. 8-1 is the resistor in circuit B. In that circuit, $X_L = R$. Assume that the switches have been closed for the amount of time needed to allow the currents in both circuits to reach the maximum value. Which inductor will generate the higher voltage when the switches are opened?

A. The circuit in A

B. The circuit in B

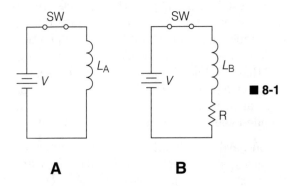

■ 8-1

A   **B**

5. Refer to the circuit in Fig. 8-2. Which of the following statements is correct?

A. The current through R3 is equal to the current through R4.

B. The current through R3 is greater than the current through R4.

C. The current through R3 is less than the current through R1.

D. There is no current through R3.

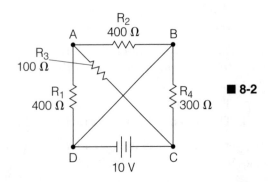

■ 8-2

6. Is the following statement correct? The voltmeter in Fig. 8-3 should indicate a voltage of 30 V.

    A. The statement is correct.

    B. The statement is not correct.

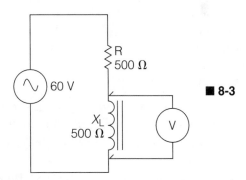

■ 8-3

7. Two capacitors are connected in series. Each capacitor has a capacitive reactance of 50 ohms. What is the capacitive reactance of the series combination? _____ ohms

8. An oscilloscope display shows five cycles of sine-wave voltage. Increase the oscilloscope sweep speed to see:

    A. more cycles.

    B. fewer cycles.

9. Which of the following is not a characteristic of an ideal op amp?

    A. Virtual ground input

    B. Rapid slew rate

    C. Common mode rejection

    D. Exponential rolloff

    E. High open-loop gain

10. A soldering iron should be tinned at:

    A. the lowest temperature possible.

    B. the highest temperature possible.

11. Which of the following bridges is used for measuring capacitance?

    A. Wheatstone bridge

    B. Maxwell bridge

    C. Schering bridge

    D. Wein bridge

12. An air gap might be added to the core of an inductor so that:

   A. it will be easier to saturate.
   B. cannot be saturated.

13. Inductive reactance is measured in:

   A. degrees.
   B. ohms.
   C. henries.
   D. darafs.
   E. farads.

14. Two identical square waves are passed through two different amplifiers. Which of the following statements is true?

   A. The one with the longer rise time has the wider frequency response.
   B. The one with the shortest rise time has the wider frequency response.
   C. Nothing can be said about the frequency response by looking at the rise time.

15. For a dc power supply that is operated from an ac power line, it is better to have:

   A. a low ripple factor.
   B. a high ripple factor.

16. A dual-trace oscilloscope provides a "chop" mode for looking at:

   A. high-frequency waveforms.
   B. low-frequency waveforms.

17. Which of the following ratings for current-measuring analog meter movements is for the more sensitive instrument?

   A. 20,000 ohms per volt
   B. 50,000 ohms per volt

18. When there are a few shorted turns in the primary winding of an isolation transformer the secondary voltage will be:

   A. slightly higher than expected.
   B. slightly lower than expected.
   C. the same as if there were no shorted turns in the primary winding.

19. The reciprocal of reactance is:

   A. conductance.
   B. measured in darafs.
   C. Choices A and B are both correct.
   D. Neither A nor B is correct.

20. The inductance of an air-core coil does not depend upon:

    A. the shape of the coil.
    B. the number of turns of wire.
    C. the coil current.
    D. the distance between the turns.

21. In which of the following components might you expect to find a Faraday shield?

    A. Power transformer
    B. Electrolytic capacitor
    C. Wire-wound resistor
    D. (Faraday shields are not used in components.)

22. Which of the following is the output in the logic circuit of Fig. 8-4?

    A. A OR B
    B. NOT (A AND B)
    C. NOT A AND NOT B
    D. A AND B

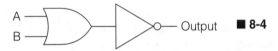

23. Which of the following are the materials used to make a solder for surface mount repair or replacement?

    A. Lead, tin, and zinc
    B. Lead, copper, and zinc
    C. Lead, tin, and silver
    D. Lead, tin, and copper

24. To decrease the bandwidth of the circuit in Fig. 8-5, move the arm of the variable resistor toward:

    A. x.
    B. y.

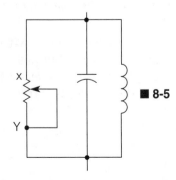

25. Is the following statement correct? *If a fuse is good, it must have a resistance of zero ohms.*

    A. The statement is correct.
    B. The statement is not correct.

26. Write the equation for the time constant (T) of an inductor in series with a resistor.

27. In the transformer symbol shown in Fig. 8-6, the broken line means:

    A. there is a direct connection between the primary and secondary windings.
    B. there is no connection between the primary and secondary windings.
    C. Neither choice is correct.

 ■ 8-6

28. What is the impedance ratio of a transformer that has a primary-to-secondary turns ratio of 11 to 5?

    A. $\dfrac{11}{5}$

    B. $\dfrac{5}{11}$

    C. $\dfrac{121}{25}$

    D. $\dfrac{25}{121}$

29. The amplifier in Fig. 8-7 is given a sawtooth frequency response test. The output display on an oscilloscope is shown on the illustration. Which of the following is the correct response to the result?

    A. The amplifier has a poor low-frequency response.
    B. The amplifier has a poor high-frequency response.
    C. The amplifier has an excellent low-frequency response.
    D. None of the choices is correct.

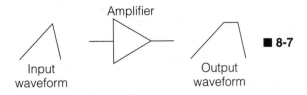

Amplifier

Input
waveform

Output
waveform

■ 8-7

30. A certain resistor has the following color code:

Blue, Purple, Orange, Red, Red

What is the highest resistance the resistor can have and still be in tolerance? _____ ohms

31. When you apply a low-frequency pure sine-wave voltage to the input of a differentiating circuit the output voltage waveform should be a:

A. sawtooth wave.
B. a square wave.
C. triangular wave.
D. sine wave.

32. What two-terminal component is represented by the connection shown in Fig. 8-8? _____

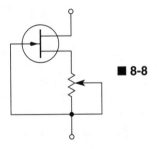

■ 8-8

33. What type of gate is represented by the following truth table?

A B Output
0 0   0
0 1   1
1 0   1
1 1   0

Answer: _____

34. Refer to Fig. 8-9. The capacitor is new and it has never been charged. What is the voltage across terminals A and D?

Answer: _____

221

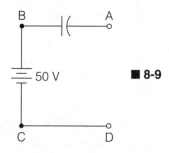

■ 8-9

35. Refer to the circuit in Fig. 8-10. The capacitor is new and it has never been charged. What is the voltage across the capacitor at the end of two time constants from the time the switch is closed?

Answer: _____

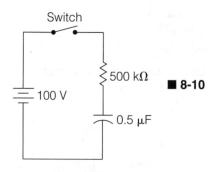

■ 8-10

36. Assume the capacitor in the circuit of Fig. 8-11 is new and it has never been charged. If a 50-kΩ resistor is connected across terminals A and B, how long after the switch is closed will it take for the capacitor (by definition) to become fully charged?

Answer: _____

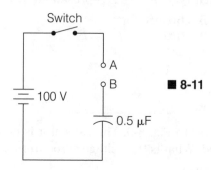

■ 8-11

37. In the circuit of Fig. 8-12, the coil has a high inductance and the circuit has been in operation for a long time. When the switch is opened the counter voltage will produce an arc across the switch contacts. That counter voltage will:

    A. send a reverse voltage across the switch that it is so high that it will produce a spark across the switch terminals.

    B. add to the battery voltage and try to keep the current flowing through the switch by producing a spark between the switch contacts.

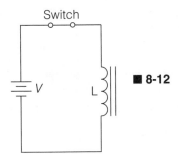

■ 8-12

38. Electric energy is measured in:

    A. watts.

    B. volt-amperes.

    C. vars.

    D. None of these choices is correct.

39. A surface mount resistor is marked 273. That means it is:

    A. shown as R273 on the schematic drawing.

    B. a precision 273-Ω resistor.

    C. it is rated at 27 Ω, 3 W.

    D. it is a 27-kΩ resistor.

    E. None of these choices is correct.

40. The highest number you can count using three flip-flops is:

    A. 3.

    B. 7.

    C. 8.

    D. 9.

41. To decrease the frequency of the relaxation oscillator shown in Fig. 8-13, move the arm of the variable resistor:

    A. toward x.

    B. toward y.

    C. Neither choice is correct because the variable resistor does not control the frequency.

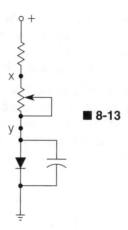

■ 8-13

42. A series of short pulses delivered to the input of an integrating circuit results in a step voltage at the output. In order to get that result, the time constant of the integrating circuit must be:

    A. long.
    B. short.

43. The circuit in Fig. 8-14 is a class-A amplifier made with a silicon transistor. The voltage at point x is 1.3 V. What voltage would you expect to measure at point y? _____V

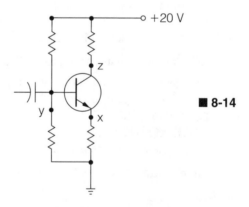

■ 8-14

44. Refer again to the class-A amplifier in Fig. 8-14. What value of voltage would you expect to measure at point z? _____V

45. Where is the energy stored in a capacitor?

    A. In the leads
    B. In the plates
    C. In the dielectric

46. Where is the energy stored in an inductor?

    A. In the leads
    B. In the conductors
    C. Neither choice is correct.

47. What is the name of the amplifier configuration in Fig. 8-15?

_____

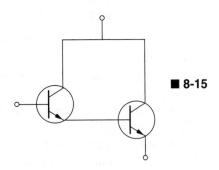

■ 8-15

48. Refer to Fig. 8-16. What is the voltage at point y with respect to point x?

    A. 5 V
    B. 4.3 V
    C. 3.3 V
    D. 0 V

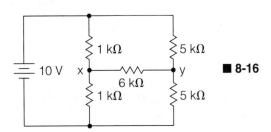

■ 8-16

49. Which of the following components is sometimes used as a parasitic suppressor?

    A. Thermistor
    B. 4-layer diode
    C. Optical coupler
    D. Ferrite bead

50. Is the following statement correct? *In a purely inductive circuit, there is no power dissipated.*

   A. Correct
   B. Not correct

51. What is another name for a beta-squared amplifier?

   A. Push-pull amplifier
   B. Complementary amplifier
   C. Quasi-complementary amplifier
   D. Darlington amplifier

52. Which of the following is not a characteristic of transformer-coupled amplifiers?

   A. Can be used to step up or step down voltage
   B. Passes ac but blocks dc
   C. Can be used to tune to pass a range of frequencies and reject all other frequencies
   D. Lowest cost of all coupling methods
   E. Can be used for impedance matching

53. If you increase the gain of an amplifier its bandwidth automatically:

   A. increases.
   B. decreases.
   C. Neither choice is correct because the gain and bandwidth of an amplifier are not related.

54. Another way of expressing transistor beta is:

   A. h_{FE}
   B. h_{FB}
   C. h_{FC}
   D. h_{FA}

55. Where in a microprocessor is simple addition and subtraction performed?

   A. ALU
   B. ACIA
   C. PIO
   D. DDR

56. Increasing the length of a coil without increasing the number of turns will:

   A. increase the inductance of the coil.
   B. decrease the inductance of the coil.
   C. not affect the inductance of the coil.

57. Is the following statement correct? *The hysteresis loss of a transformer design can be reduced by laminating the iron core.*

    A. The statement is correct.
    B. The statement is not correct.

58. Which of the following is most similar to NEPERS:

    A. SOMES
    B. PHONS
    C. DECIBELS
    D. Centimeters per square inch

59. Figure 8-17 shows the leading edge of a square wave. What is the approximate rise time of the display if the horizontal axis is calibrated for 15 microseconds per division.

    Answer: _____ microseconds

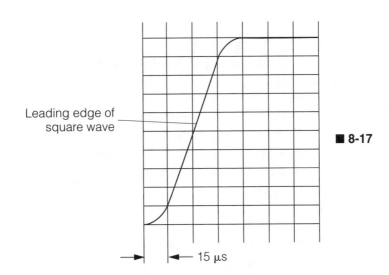

Leading edge of
square wave

■ 8-17

15 μs

60. Is the following statement correct? *In a superheterodyne receiver, the local oscillator must always be operated at a frequency that is higher than the frequency of the signal being received.*

    A. The statement is correct.
    B. The statement is not correct.

61. Which of the following is not a diode?

    A. Esaki

    B. Schottky

    C. Magnetron

    D. Ruby Laser

62. Assume that ϕ is the phase angle between the voltage and current in an ac circuit. The power factor is:

    A. $\sin \phi$

    B. $\cos \phi$

    C. $\tan \phi$

    D. cotangent ϕ

63. It takes exactly 25 ms for a certain oscilloscope to produce one line of trace. A sine-wave voltage is delivered to the vertical input terminals of the 'scope. What frequency must the input sine-wave voltage have to produce four complete cycles of display?

    A. 10 Hz

    B. 40 Hz

    C. 120 Hz

    D. 160 Hz

64. Which of the following would be found in a phase locked loop?

    A. BFO

    B. MO

    C. VCO

    D. None of the above choices is correct.

65. In a superheterodyne receiver that has a diode detector, a CW signal can be demodulated with the use of a:

    A. VCO.

    B. CPO.

    C. BFO.

    D. None of the above choices is correct.

66. In the circuit of Fig. 8-18, resistor R is connected as a:

    A. rheostat.

    B. potentiometer.

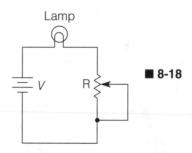

■ 8-18

67. The first three colors on a color-coded resistor are: yellow, purple, and silver. What is the resistance of the resistor?

_____ ohms.

68. Refer to the circuit in Fig. 8-19. What is the maximum amount of power that can be dissipated by R?

    A. 0.06 watts

    B. 0.6 watts

    C. 6.0 watts

    D. None of these choices is correct.

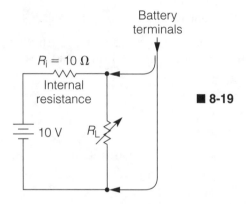

■ 8-19

69. To transmit and receive using the same antenna use a:

    A. diplexer.

    B. duplexer.

70. Disregarding the very strange audio specifications, when you multiply RMS voltage by RMS current, you get:

    A. RMS power.

    B. average power.

    C. peak power.

    D. peak-to-peak power.

71. Name the three main busses in a microprocessor.

_____

_____

_____

72. In a class-A audio power amplifier, an electrolytic capacitor might be connected across the emitter resistor to:

    A. increase the gain.

    B. increase the bandwidth.

73. In a certain transformer, the turns ratio ($N_P:N_S$) is 3 to 1. If a 5-ampere current is flowing in the primary winding, how much current is available in the secondary winding?

    A. 0.6 ampere

    B. 5 amperes

    C. 15 amperes

    D. None of the choices is correct.

74. What is the ripple frequency output of a bridge rectifier if the frequency of the input power is 400 hertz?

    A. 200 hertz

    B. 400 hertz

    C. 600 hertz

    D. 800 hertz

75. Conductance is measured in:

    A. vars.

    B. mhos.

    C. siemens.

    D. darafs.

230

Answers to practice questions

1. Choice C is correct. The component is sometimes called a *Hall sensor*. Figure 8-20 shows how it operates. It consists of a slab of semiconductor material with current flowing through it. In the absence of an external field, the current distributes itself evenly across the device. Because electrons repel each other, it is natural for them to disburse (as shown in the illustration).

When a transverse magnetic field passes through the semiconductor slab, the current flow is rearranged. The magnetic field causes the electrons to crowd against one side of the slab, producing a negative voltage at that side. The other side has a deficiency of electrons and it becomes a positive electrode. The amount of voltage generated depends directly upon the strength of the magnetic field.

Hall devices are used as magnetic sensors. One example is an automatic speed control for an electric motor. Every speed control starts with a measurement that shows how fast the motor is turn-

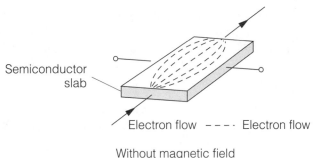

Semiconductor slab

Electron flow - - - - Electron flow

Without magnetic field

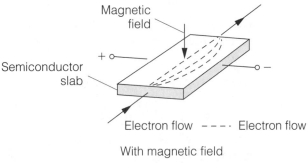

Magnetic field

Semiconductor slab

Electron flow - - - - Electron flow

With magnetic field

■ 8-20

ing. Without that measurement, the automatic circuitry would not "know" whether the speed needs to be increased or decreased. The Hall device is located near a rotating part of the motor output shaft where magnets are embedded.

Each time a magnet passes the Hall device, a voltage pulse is generated. The number of pulses per second is compared with the pulse frequency from an oscillator. That oscillator frequency represents the desired speed. From the difference between the pulse frequency from the Hall device and the oscillator frequency, the electronic circuitry increases or decreases the speed of the motor (if necessary).

2. Choice A is correct. There are three classifications of magnetic materials: ferromagnetic, paramagnetic, and diamagnetic.

Ferromagnetic materials are strongly attracted to a magnetic field. The name comes from the word *ferrous*, which means *iron*. However, not all ferromagnetic materials are made with iron. One example is Alnico. It is made with aluminum, nickel, and cobalt. Yet, it has very strong magnetic properties. Alnico is used as a permanent magnet in PM speakers.

The second group is called *paramagnetic*. Theoretically, it is a material that is not affected by a magnetic field. In real life, there are very few materials that are paramagnetic. In a paramagnetic material, the influence of a magnetic field is so slight that it can be disregarded.

The third group is *diamagnetic*. A diamagnetic material is repelled by a magnetic field. A thin copper sphere is an example of a diamagnetic material.

3. Choice D is correct. Stepping motors have two advantages that make them very useful in some electronic applications. They are:

☐ Their shafts can be accurately positioned by using a specific number of pulses. This feature makes them very useful in robotics.

☐ Their speed can be accurately controlled using digital technology. For example, the motors in VCRs are operated at precise speeds and stepping motors are used for this purpose.

4. Choice B is correct. Remember that when the switch is opened, the collapsing magnetic field around the coil induces a voltage that tries to keep the current from changing. In this case, the current is trying to stop so that the induced voltage will add to the battery voltage to keep the current going. That is what produces an arc across the switch.

With the resistor in the circuit, the induced voltage must be higher, and therefore, the arc across the switch will be more intense. In other words, the arc will be made up of a higher current. This is the reason that resistors are sometimes added in series with spark plugs in a car engine.

5. Choice D is correct. Study the circuit carefully. Observe that there is a short circuit between points D and B. With that short circuit in place, there is no current flowing through resistors R1, R2, and R3. All of the circuit current flows through R4. Understand that point A in the circuit is at zero volts because it is connected to a point where there is no current. There is no current through R3.

6. Choice B is correct. Observe that the circuit has a resistance and inductive reactance in series. The same current flows through the two components and the voltages must be in quadrature. That is another way of saying that the effect of the voltage across the resistor is at a right angle to the voltage across the inductor (Fig. 8-21).

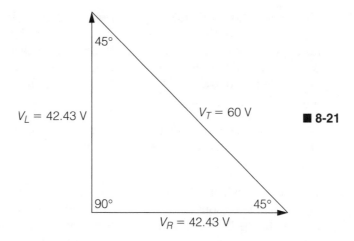

From basic geometry, the two legs of the right triangle must be equal and the angle 45 degrees. The voltage across each component is calculated as: when V is the applied voltage, V_R is the voltage across the resistor, and V_L is the voltage across the inductor:

$$V^2 = V_R^2 + V_L^2$$

$$\text{Because } V_R = V_L, \text{ let } V_R^2 + V_L^2 = 2V_X^2$$

$$60^2 = 2V_X^2$$

$$V_X^2 = 1800$$

$$V_X = 42.43 \text{ volts} = V_R = V_L$$

Therefore, the voltage across the resistor is 42.43 V and the voltage across the inductive reactance is 42.43 V when the applied voltage is 60 V. (Note: The numbers have been rounded off to simplify the calculations.)

7. The correct answer is 100 ohms. If you put 25 ohms as the answer, you are confusing capacitances in series with capacitive reactances in series. Series capacitances divide, but series reactances add.

8. Choice B is correct. Suppose, for example, that the speed is five times faster. That means it only takes one-fifth of the time for the beam to move across the oscilloscope screen. With the sweep at one-fifth of the time, there is only enough time to see one cycle.

By way of contrast, if the sweep speed is decreased, it takes a longer time for the beam to move across the screen and there would be more time for displaying cycles.

233

9. Choice D is correct. An ideal operational amplifier must have a linear rolloff. In the benchmark op amp (a 741), the linear rolloff makes it an easy matter to calculate the bandwidth and gain. In some op amps, like the switching type, the rolloff is not perfectly linear, but that disadvantage is offset by the fact that the op amp is more efficient.

10. Choice A is correct. The reason for tinning the soldering iron at the lowest possible temperature is that there is less corrosion produced in the tinning process.

11. Choice C is correct. Each of the bridges listed as choices have practical applications. The Wheatstone bridge is used for measuring resistance. The Maxwell bridge is an easily constructed bridge used for measuring inductance. The Shearing bridge measures capacitance. The Wein bridge is used for measuring frequency. These bridges are used as laboratory instruments where very precise measurements are needed.

12. Choice B is correct. Remember that it is not possible to magnetically saturate air. When iron becomes saturated, it means that no increase in flux can occur even though there is an increase in the applied magnetic force. No amount of magnetomotive force can saturate air.

13. Choice B is correct. This might seem like a very easy question, but you would be surprised at the number of experienced technicians who answer "henries."

14. Choice B is correct. The question describes the square wave test for determining a frequency response of an amplifier. As a general rule this is used as a qualitative test. In other words, a specific bandwidth is not being indicated as the test is usually performed. Think of this. The fast rise time (or, the shortest rise time) has to occur when the amplifier is able to change from a minimum to maximum output voltage very rapidly. That is also true of an amplifier when it is passing a very high frequency. With a high frequency, the voltage changes from minimum to maximum very rapidly so the rise time is a good test of the high-frequency response. Figure 8-22 shows two cases of square-wave outputs. One shows poor high-frequency response and the other shows poor low-frequency amplifier response. Of course, if the output is a perfect square wave then the amplifier has a very broad frequency response.

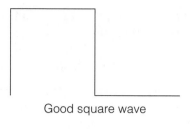

Good square wave

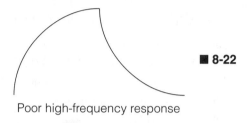

■ 8-22

Poor high-frequency response

Poor low-frequency response

15. Choice A is correct. Here is an equation often given for ripple factor:

$$Ripple\ factor = \sqrt{\frac{I_{rms}}{I_{dc}} - 1}$$

You might also see it expressed as the ratio of the RMS value divided by the average value of current. In either case, the lowest value is the best value. You will not be asked to calculate the ripple factor in a CET test, but keep in mind the fact that you are looking for the lowest possible value.

16. Choice B is correct. In the chop mode, the beam is moved back and forth rapidly between the two displays. Therefore, each display is chopped into very small pieces (which might look continuous to your eye). That is the ideal mode for looking at waveforms that have a very low frequency.

17. Choice B is correct. The sensitivity of an analog meter movement is the reciprocal of the ohms-per-volt rating. That gives the amount of current that is required for full-scale deflection.

For 20,000 ohms per volt, a full-scale deflection requires 50 microamperes. For 50,000 ohms-per-volt, full-scale deflection occurs with only 20 microamperes. Therefore, the 50,000 ohms-per-volt rating is for the more sensitive instrument.

18. Choice A is correct. An isolation transformer has a primary-to-secondary turns ratio of one to one. If there are a few turns shorted in the primary, it means effectively that there are less turns in the primary than in the secondary and the transformer becomes a step-up type. That results in a higher secondary voltage.

19. Choice D is correct. The reciprocal of reactance is susceptance. Be sure you know the following reciprocals:

☐ The reciprocal of impedance is admittance.

☐ The reciprocal of reactance is susceptance.

☐ The reciprocal of resistance is conductance.

20. Choice C is correct. The factors that determine inductance are the number of turns of wire and the shape of the coil. The coil radius and distance between the turns (length) determine the shape of the coil.

21. Choice A is correct. A Faraday shield is a grounded screen between the primary and secondary windings of a power transformer. Its purpose is to prevent capacitive coupling between the primary and secondary. Remember that the primary and secondary windings are made with conductors separated by insulation. Any time you have conductors separated by insulation, you have a capacitor.

The capacitance is very low and not a serious problem at power line frequencies. However, transient (spike) voltages are made up of very high frequencies and they would normally pass through the primary-to-secondary winding capacitance. Hence, the Faraday shield is primarily used to prevent transients from coupling into the transformer secondary circuit.

Transient voltages that arrive on the power transformer primary (by way of the ac power line) can have a very high amplitude that can destroy secondary circuit components.

The Faraday shield connection usually comes out of the transformer along with other wires. You can recognize it by the fact that it is usually a braided wire. That wire must be grounded if it is to be effective in eliminating transients in the secondary.

22. Choice C is correct. NOT A AND NOT B ($\overline{A}\ \overline{B}$) The output is obviously NOT A OR B. However, DeMorgan's theorem states that you can break the bar and change the sign. Refer to Fig. 8-23, which shows two applications of the DeMorgan theorem.

Break the bar and change the sign

$$\overline{A + B} = \overline{A}\ \overline{B}$$

■ 8-23

$$\overline{AB} = \overline{A} + \overline{B}$$

23. Choice C is correct. This choice of materials for the solder makes it possible to solder at a lower temperature. The small surface-mount devices can be easily destroyed by high temperatures. So, the proper surface-mount soldering process requires fast operation.

24. Choice B is correct. Think of it this way: If you move the arm all the way to point x, there will be a short circuit across the tuned circuit. In other words, it is like connecting the wire across the tuned circuit and the bandwidth of a piece of wire is enormously wide. As you move the arm toward y the bandwidth gets narrower.

25. Choice B is correct. If a fuse *really* had a resistance of zero ohms, then power dissipated by the fuse wire would be zero watts and the fuse could not burn out. Never forget that the fuse burns out because current flowing through the resistance of the fuse produces heat. When that heat is at the melting point, the fuse element melts and the fuse is "blown."

26.

$$T = \frac{L}{R}$$

where: T is the time constant in seconds
L is the inductance in henries
R is the resistance in ohms

27. Choice C is correct. The broken line represents the Faraday shield that was covered in question 21. The physical appearance of this shield connection is usually a braided wire. We repeat: The braided wire must be grounded in order for the shield to be effective!

28. Choice C is correct. You might see the equation for transformer impedance in two different ways:

$$\left[\frac{N_P}{N_S}\right]^2 = \frac{Z_P}{Z_S}$$

or

$$\frac{N_P}{N_S} = \sqrt{\frac{Z_P}{Z_S}}$$

Substituting the turns ratio into the first equation gives:

$$\left(\frac{N_P}{N_S}\right)^2 = \frac{(N_P)^2}{(N_S)^2} = \frac{121}{25} = \frac{Z_P}{Z_S}$$

29. Choice D is correct. The waveform shown in the illustration represents clipping. That can be caused by either of two causes: excessive input drive or insufficient dynamic range of the amplifier. Refer to Fig. 8-24, which shows sawtooth outputs for various amplifier conditions.

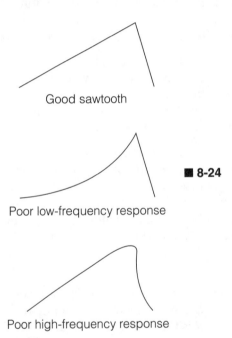

Good sawtooth

■ 8-24

Poor low-frequency response

Poor high-frequency response

The sawtooth amplifier test gives the same information as the square-wave test. However, it is difficult to see if there is clipping when you use the square-wave test. Some technicians use both tests. Those tests are usually qualitative. In other words, the tests are not usually used to determine the bandwidth of an amplifier as a range of frequencies in numbers. Only the high-frequency response and the low-frequency response are indicated using the quick and easy square-wave and sawtooth- wave tests. However, remember that an approximate quantitative test (giving actual numbers) can be added by measuring the rise time of the square wave.

238

The equation relating rise time and approximate bandwidth is given here:

$$Bandwidth = \frac{0.35}{Rise\ time}$$

Don't be misled by the fact that an equation is given for finding the bandwidth from the risetime. At best, this equation will give you an estimate of bandwidth. It is an "empirical" equation. In other words, it was "derived" by taking a lot of actual measurements and fitting them into an equation.

The best method of measuring actual bandwidth is to plot the range of frequencies vs. the amplitude for each frequency. Actually, not every frequency is plotted. Instead selected frequencies are used.

The second best method of measuring bandwidth is to use a sweep generator and oscilloscope. Specific frequencies can be marked on the bandwidth display using a marker generator.

You might be asked about bandwidth measurements in any level of CET test.

30. The correct answer is 68,340 ohms. It is a 67-kΩ resistor with a ±2% tolerance. The last red stripe in the color code is a reliability code, which is not normally used by technicians. It shows how many resistors must pass inspection out of a given number of resistors purchased.

239

31. Choice D is correct. The differentiating circuit passes the low-frequency sine-wave without modification (Fig. 8-25).

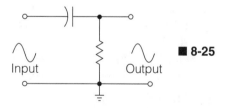

Input   Output   ■ 8-25

32. The correct answer is constant current diode. If you purchase a constant-current diode, it might be made with the connection shown in Fig. 8-8. This is one component that can be easily fabricated, but you have to adjust the resistor to get the constant current you need.

33. The correct answer is exclusive OR (by definition). You should make it a point to learn the truth tables for the following

Practice taking CET tests

gates: AND, OR, NOT (inverter), NAND, NOR, exclusive OR, and, logic comparator.

34. The correct answer is 50 volts. A surprising number of experienced technicians miss this question. That is probably because of the following model that is sometimes used for capacitors: "A capacitor will pass ac, but block dc." That model does not apply to many applications, including this case.

Think of it this way: If there is no charge on the capacitor, there is no voltage across the capacitor. Therefore, the voltage at point A must be identical to the voltage at point B. In other words, there is a 50-volt potential between points A and D.

35. The correct answer is 86.3 V. It is important to remember that the capacitor will first charge to 63% of the applied voltage in one time constant. Then, it will charge to 63% of what is remaining during the next time constant.

At the end of the first time constant:

$$0.63 \times 100 \text{ V} = 63 \text{ V}$$

During the next time constant, the capacitor will charge to 63% of the remaining voltage. The remaining voltage is:

$$100 - 63 = 37 \text{ V}$$

63% of that remaining voltage is:

$$0.63 \times 37 = 23.3 \text{ V}$$

Add the 23.3 volts to the 63 volts attained during the first time constant and you get:

$$63 + 23.3 = 86.3 \text{ volts (answer)}$$

Using this reasoning, you can determine the voltage across a capacitor in an RC time constant circuit for any number of time constants. You don't need any special time constant equations for this type of problem, provided that you are only interested in the voltage at some whole-number time constant.

You can get better accuracy if you use 63.2 as the time constant. You can also use the same reasoning if you want to find the voltage across a capacitor in an RC time-constant discharging circuit. All you have to remember is that the capacitor will discharge to 37% of its maximum voltage; that is, to 37% of the starting voltage across the capacitor. If you want better accuracy, use 36.8%.

At the second time constant, the capacitor will discharge another 37%, and another 37% during the next time constant, and so on for any number of time constants.

36. By definition, the time required for full charge is five time constants. The calculation is shown here:

$$\text{One time constant, } T = R \times C$$

$$= (50 \times 10^3) \times (0.05 \times 10^{-6})$$

$$T = 2.5 \text{ milliseconds}$$

Full charge requires 5 time constants:

$$Full\ charge = 5T = 5 \times 2.5 \text{ ms} = 12.5 \text{ milliseconds}$$

37. Choice B is correct. When the switch is opened, the counter voltage will add to the applied voltage to keep the current moving through the switch. That is what causes the arc across the switch when it is opened.

The arc can be destructive to the switch contacts. For that reason, you are likely to see a voltage-dependent resistor (VDR) or capacitor across the switch. The purpose of those components is to suppress the arc and prolong the lifetime of the switch.

38. Choice D is correct. Energy is the amount of power exerted over a period of time. For example, one watt dissipated over a period of one second is called a *watt-second* or *Joule*. A *kilowatt-hour* is a kilowatt dissipated over a period of one hour.

It is interesting to note that you pay the "power" company for energy, not for power in watts. A watt-hour meter is used for measuring the energy that you use each month or, for whatever time period the power company uses for billing. In the proposed newer installations, it will not be necessary for a meter reader to come to your house. The number of watt-hours can be measured at the power plant.

It is good to remember that power is the time rate of using energy.

$$Power = Energy/Time$$

39. Choice D is correct. The first two numbers refer to the first two digits of resistance. The third number tells the number of zeros.

40. Choice B is correct. With three flip-flops, you can make $(2^3) =$ eight counts. However, the first count is zero, so the eight counts are 0 through 7.

41. Choice A is correct. Moving the arm of the variable resistor toward point y increases the resistance in the RC time constant of the circuit. As with most relaxation oscillators, the frequency depends upon the RC time constant. Increasing R increases the time constant; therefore, it increases the time for one cycle. That, in turn, decreases the frequency.

241

42. Choice A is correct. With a long time constant, the capacitor holds the voltage produced by a pulse until the next pulse comes along. Each pulse adds to the voltage already stored and that is how you get the step voltage.

43. The correct answer is about 2 volts. Remember that with a silicon transistor, about 0.7 V is between the emitter and base. That voltage is added to the 1.3 volts at point x, giving a total of (about) 2.0 volts.

The reason that we say "about 2 volts" is that the emitter-to-base voltage of silicon transistors is sometimes given as 0.6 volts; at other times, it is given as 0.7 volts. You should remember that those voltages are only for low-power voltage amplifiers. With a silicon power transistor, the voltage can be as high as 1.2 volts!

44. The correct answer is about 10 volts. Remember that in a class-A amplifier, the collector voltage is usually about one-half the power supply voltage. That is not meant to be an exact value, but it is important for troubleshooting.

If an amplifier is working properly, a voltage measurement at the collector should show about half the power supply voltage or 10 volts. If that voltage is present at the collector, it indicates that the amplifier is working properly and you can go to the next stage when you are troubleshooting.

45. Choice C is correct. A surprising number of technicians have missed this question on a CET test. The energy is stored in the dielectric, and the charge is stored in the plates. To show how important the dielectric is, remember that capacitors are usually described by the material used for their dielectric. Some examples are: electrolytic, plastic, mica, etc.

46. Choice C is correct. The energy is stored in the magnetic field surrounding the inductor.

47. The configuration is called a *Darlington amplifier*. It has a high-input impedance and high power gain. The disadvantage of the Darlington amplifier is that it produces high internal heat. The Darlington amplifier is also known as a *beta-squared amplifier* because if the betas of the two transistors are identical, the gain is beta × beta or beta^2.

48. Choice D is correct. You can think of the circuit as being a balanced Wheatstone bridge. When a wheatstone bridge is balanced, the center leg (between points x and y) has no current. A microammeter between those points would display zero microam-

peres of current, so the center leg can be removed without changing the balance.

49. Choice D is correct. A ferrite bead is mounted on the wire lead from the input to an amplifier (Fig. 8-26). It behaves as if an inductor was connected in series with that lead.

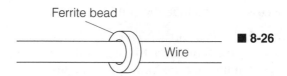
Ferrite bead

Wire

■ 8-26

Parasitics are undesired frequencies that are generated in high-gain amplifiers and thyristors. High-frequency parasitics can be destructive in a circuit and the ferrite bead is used to prevent the oscillation from occurring.

It is not unusual to see as many as three or four ferrite beads on one lead. They behave like inductors in series. In some circuits, resistors or inductors are also used to eliminate parasitic oscillations. Remember that amplifiers with a very high gain are likely sources of parasitic oscillations.

50. Choice A is correct. In a purely inductive circuit (that is, a circuit with no resistance), no power is dissipated. Likewise, in a purely capacitive circuit, no power is dissipated. In an ac circuit, the power is dissipated in resistance.

51. Choice D is correct. This was covered in the previous question on Darlington amplifiers. See question and answer #47.

52. Choice D is correct. This is not the cheapest method of coupling amplifiers. However, the advantages sometimes make the extra costs worthwhile. For example, the transformer can be tuned to pass a range of frequencies and reject all other frequencies. As another example, the transformer can be used for impedance matching.

53. Choice B is correct. In *ES&T* magazine, a prize was once offered to any technician who could increase the gain of a given amplifier without decreasing its bandwidth. There were a number of tries, but none of them worked because gain and bandwidth are tradeoffs in an amplifier.

54. Choice A is correct. The letters stand for the *hybrid Forward Emitter* characteristic. If the subscripts are lowercase, the parameter is for ac or signal conditions.

243

55. Choice A is correct. The letters stand for *Arithmetic Logic Unit*.

56. Choice B is correct. Think of it this way. If you stretched out the coil, it would become a straight piece of wire and its inductance would be negligible, except at very, very high frequencies. If those high frequencies were to be considered, it would have been necessary to include the condition in the question.

57. Choice B is correct. The only way you can decrease hysteresis loss in a transformer is to use a better type of iron for making the core. Laminating the iron core is the method used to reduce eddy current losses.

58. Choice C is correct. In the United Sates, *decibels* are commonly used to express ratios of power, but *nepers* are used in other countries. Decibels are calculated using logarithms (Log) to the base 10. Nepers are calculated using logarithms (Ln) to the base epsilon. *Somes* and *phons* are units of measurement in audio.

59. The correct answer is a little over 30 microseconds. The rise time is the time that it takes to go from 10% to 90% of the maximum amplitude.

60. Choice B is correct. In multiband receivers, the local oscillator frequency might be above the incoming frequency for one band and below the incoming frequency for an adjacent band.

61. Choice D is correct. All other choices are names of diodes.

62. Choice B is correct. The power factor is equal to the cosine of the phase angle by definition. Power factor is a measure of how closely the voltage and current are in phase. A power factor of 1.0 corresponds to a zero-degree phase angle because the cosine of zero is 1.0.

63. Choice D is correct. The time for one sweep is 25 milliseconds. The reciprocal of that time gives a frequency of 40 hertz. Therefore, if the frequency is 40 hertz, one sine wave would be displayed. To get four complete sine waves, the frequency has to be four times higher than 40 hertz (160 hertz).

64. Choice C is correct. The letters stand for *voltage controlled oscillator*. It is very important for you to review the characteristics of the phase-locked loop before taking any CET Test. *BFO* stands for *beat frequency oscillator* and *MO* stands for *master oscillator*.

65. Choice C is correct. The letters *CW* stand for *continuous wave* transmission. An example of CW transmission is transmission by Morse code. The letters *BFO* stand for *beat frequency oscillator*. Think of it this way. The purpose of the diode detector is to separate audio and IF frequencies. If there is no audio, there is no detector output. In order to get an output, an oscillator signal must be added to heterodyne with the IF frequency. Questions about radio circuits are more likely to be found in the Journeyman Consumer Electronics Option.

66. Choice A is correct. Remember that a rheostat controls the current in a circuit and a potentiometer controls the voltage in a circuit. In Fig. 8-18, adjusting the variable resistor changes the amount of current that flows through the lamp, so the variable resistor is being used as a rheostat.

67. The correct answer is 0.47 ohms. If the third band had been gold, it would be a 4.7-ohm resistor. Color codes for resistances less than one ohm are frequently missed by technicians taking a CET test. The color codes for 10-ohm and 100-ohm resistors are also frequently missed by technicians taking CET tests.

68. Choice D is correct. The maximum amount of power that can be dissipated in a dc circuit occurs when the load resistance is equal to the internal resistance of the source. In this case, the maximum output power occurs when the load resistor (R_L) equals 10 ohms. Under that condition, half of the total power dissipated in the circuit is dissipated in the load resistor.

$$Total\ power = \frac{V^2}{R} = \frac{100}{20} = 5\ W$$

Half of that total power is dissipated in the load resistor. So, the maximum power that can be delivered to the load resistor in the circuit of Fig. 8-19 is 5/2 or 2.5 watts.

69. Choice B is correct. A diplexer permits two different frequencies to be received for two different receivers. The duplexer permits transmitting and receiving with a single antenna.

70. Choice B is correct. Somewhere along the line someone who knows absolutely nothing about electricity decided that RMS voltage times RMS current should give RMS power. The terminology (RMS power) is absurd. You will only find the term *RMS power* used in reference to audio systems. It seems that once it got started, no one seemed to be able to know how to stop it.

71. ☐ Data bus

☐ Control bus

☐ Memory bus

72. Choice A is correct. An electrolytic capacitor across the emitter resistor of an audio power amplifier is used to eliminate degeneration. That, in turn, increases the amplifier gain and decreases its bandwidth. Designers usually prefer to have the amplifier constructed without the capacitor. They feel that the broader bandwidth is preferable to the higher gain. Therefore, in many low-cost radios, that emitter bypass capacitor is purposely omitted.

73. Choice C is correct. Remember that in a step-down transformer the available current in the secondary winding is increased. The transformer current equation is:

$$\frac{N_P}{N_S} = \frac{I_S}{I_P}$$

That doesn't mean that the higher current will necessarily flow in the secondary circuit. The amount of current that actually flows depends upon the resistance connected across the secondary winding. In other words, not all of the available current will flow in the secondary.

An important thing about this question is that it shows why you cannot determine the turns ratio of a transformer by measuring the primary and secondary winding resistances. Obviously, in the step-down transformer of this question, the current-carrying capacity of the secondary must be greater than for the primary. Another way of saying that is that the secondary winding must be made with wire that has a larger diameter. That makes the resistance measurements of the two windings useless for determining turns ratio.

74. Choice D is correct. A bridge rectifier is an example of a full-wave rectifier. The output ripple frequency is always twice the input frequency of the power-line in full-wave (single phase) rectifiers.

75. Choice C is correct. If you answered B (mhos), you are not keeping up with the times. The *mho*, which is *ohm* spelled backwards, was a unit of measurement for conductance for many years. It has been replaced by *siemens*. Certified Electronic Technicians must keep up with the times. Don't forget that you are never finished with studying as long as you are working in the field of electronics.

246

A time-limit test

9

ONCE, YOU COULD TAKE AN UNLIMITED AMOUNT OF TIME TO take A CET test. As always seems to happen, some people abused that privilege. It got so bad that some administrators probably wondered if they would ever see their families again.

So, the rule has been changed. Now you must complete the test within two hours. That is 1¼ minutes for each question. When you take much less time to answer easier questions, you can take minutes to answer difficult questions. If you know your subject, there just isn't that much to think about when you are answering a question.

One reason for taking the full two hours is that some people like to look over the test before handing in their paper. That can be done with the allotted time. However, you can ruin a good answer sheet by going over and over and over the test. You start seeing things that aren't really there. You start looking for exceptions to each answer.

There are no trick questions. To some people, a trick question is one they can't answer. Actually, a trick question is one that switches the meanings of terms. Here is a good example of a trick question:

"As of today, how many states are there in the United States?"

If you give a quick answer you will say that there are 50 states in the U.S. That answer would be wrong! Four of the so-called states are not really states, they are commonwealths! Don't you feel like you have been tricked?

You will *not* find that type of question in the CET test. The questions are designed to test your knowledge of the subject. Be sure you understand this: If you can't match your knowledge to one of the choices in a question, it is not because it is a trick question. Look for a different reason!

You will find more emphasis on logic and digital subjects in this test. Here are some of the things you must know in order to answer questions on digital electronics.

☐ The basic gates, their symbols and truth tables

☐ The rules of Boolean algebra

☐ Conversion of numbers from one radix to another. For example, convert the decimal number 22 (22_{10}) to a binary number (XXXXX_2).

Use the same setting to take this test as you used for the first test in this chapter. In other words, do not allow any distractions. However, place a two-hour time limit for taking the test. You don't have to take the complete two hours, but do not allow yourself one minute more.

By passing the tests in this chapter, it doesn't necessarily mean that you will pass an actual CET test, but it is a good indicator. If you can't pass the tests in this book you should devote more time to studying before you take the actual CET test. Take special note of the questions that you missed.

Practice (timed) test #2

1. Which of the following color codes is correct for a 100-ohm resistor?

    A. Brown, black, red
    B. Brown, black, brown
    C. Brown, black, black
    D. Brown, brown, black

2. The power dissipated by the resistor in Fig. 9-1 is:

    A. 40 watts.
    B. 4 watts.
    C. 8 watts.
    D. 0.8 watts.

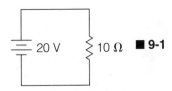
20 V   10 Ω   ■ 9-1

3. What is the battery voltage in the circuit of Fig. 9-2?

A. 40 V

B. 400 V

C. 100 V

D. 10 V

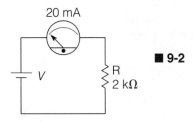

■ 9-2

4. The discharge path for C, in the circuit of Fig. 9-3, is shown by the arrow marked:

A. a.

B. b.

C. c.

D. None of these choices is correct.

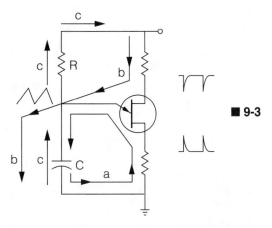

■ 9-3

5. A certain time constant circuit is made with a resistor in series with a capacitor. The capacitor has a temperature coefficient of N750. When the room temperature decreases, the time constant of the series circuit will:

A. increase.

B. decrease.

6. What is the time constant of a circuit that has a 250-millihenry coil connected in series with a 250-ohm resistor?

    A. One microsecond
    B. 62.5 milliseconds
    C. 625 milliseconds
    D. None of these choices is correct.

7. For the transformer shown in Fig. 9-4, the higher secondary voltage will occur when the switch is in the position marked:

    A. x.
    B. y.

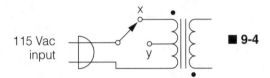

115 Vac input — ■ 9-4

8. The resistance in the circuit of Fig. 9-5 is too high. This means that the current is:

    A. too low.
    B. too high.

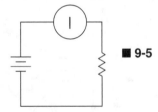

■ 9-5

9. Refer to Fig. 9-6. The applied voltage (V) in this circuit must be:

    A. 4.13 volts.
    B. 8.31 volts.
    C. 11.66 volts.
    D. 20 volts.

10. For the circuit in Fig. 9-7, find the voltage at point x with respect to the common point. The value of voltage is:

    A. 6 V.
    B. 14 V.
    C. 15 V.
    D. 16 V.

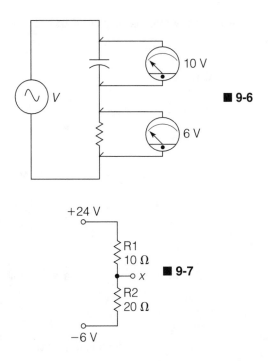

9-6

9-7

+24 V

R1
10 Ω

x

R2
20 Ω

−6 V

11. In the circuit of Fig. 9-8, what value of R_L will receive maximum power from the battery and its internal resistance (R_i)?

    A. 60 ohms
    B. 84 ohms
    C. 120 ohms
    D. Infinity

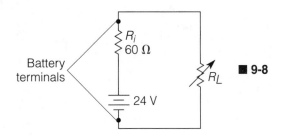

9-8

12. What is the value of the maximum power delivered to R_L in the circuit of Fig. 9-8? _____ watts

    A. 4.6
    B. 1.2
    C. 2.4
    D. 8.4

13. For the NPN transistor circuit in Fig. 9-9, the polarity at point x should be:

    A. negative.
    B. positive.

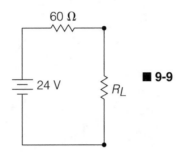

■ 9-9

14. What is the value of collector current for the circuit of Fig. 9-10?

    A. 0.982 A
    B. 0.0982 A
    C. 0.00982 A
    D. None of these choices is correct.

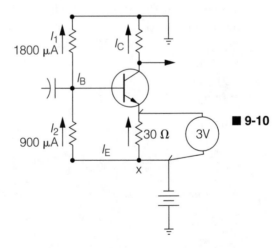

■ 9-10

15. What is the frequency of the signal displayed in Fig. 9-11?

    A. 1500 hertz
    B. 2000 hertz
    C. 2500 hertz
    D. 250 hertz

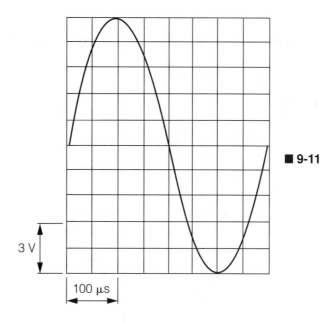

■ 9-11

3 V

100 μs

16. The RMS value of the voltage displayed in Fig. 9-11 is:
    A. 5.0 V.
    B. 5.3 V.
    C. 5.6 V.
    D. 6.0 V.

17. In a certain series-tuned circuit, the value of inductance is increased by pushing its turns of wire closer together. This will:
    A. increase the resonant frequency.
    B. decrease the resonant frequency.

18. In a PNP transistor, the base must be:
    A. positive with respect to the collector.
    B. negative with respect to the collector.

19. Diodes are connected in series as shown in Fig. 9-12 in order to:
    A. increase their current ability.
    B. increase their peak inverse voltage rating.

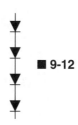

■ 9-12

253

20. In the circuit of Fig. 9-13 an electrolytic capacitor is used for coupling the signal between Q_1 and Q_2. Which of the following statements is true?

    A. Leakage of that electrolytic capacitor will certainly destroy Q_1, so it should not be used.

    B. Leakage of the electrolytic capacitor will certainly destroy Q_2, so it should not be used.

    C. Electrolytic capacitors will not pass the higher audio frequencies. Therefore, the capacitor could not be used in an audio system.

    D. The electrolytic capacitor is ok to use and its use is made possible by the low operating voltages of bipolar transistors.

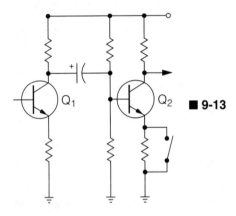

■ 9-13

21. Refer to the symbols in Fig. 9-14. Which of these symbols is for a P-channel JFET?

    A. The symbol shown in A.

    B. The symbol shown in B.

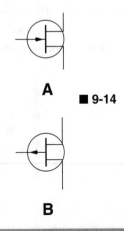

A

■ 9-14

B

22. Consider the two illustrations in Fig. 9-15. Which of the following statements is true?

    A. Neither circuit will work properly.

    B. Both circuits will work properly.

    C. The circuit in A will work properly but not the one in B.

    D. The circuit in B will work properly but not the one in A.

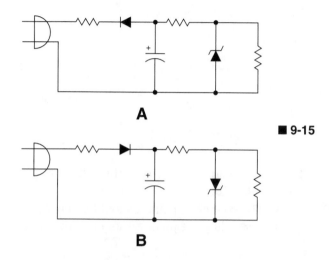

A

■ **9-15**

B

23. Consider the half-wave rectifier circuit of Fig. 9-16. What is the purpose of R1 in this circuit?

    A. Resistor R1 is used to improve the filtering of the rectifier output.

    B. Resistor R1 is used to prevent burnout of the diode during the charging of the filter capacitors.

    C. Resistor R1 is incorrectly drawn in this circuit. It should be on the right side of the diode.

    D. Resistor R1 has no purpose in this circuit and should be eliminated for improved performance.

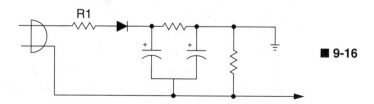

R1

■ **9-16**

24. Consider the circuit of Fig. 9-17. Which of the following statements is correct?

    A. The output signal is in phase with the input signal.

    B. The output signal is 180 degrees out of phased with the input signal.

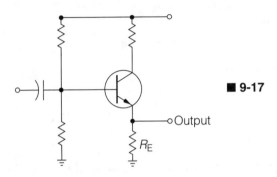

■ 9-17

25. What should the polarity of the voltage at point x be in the circuit of Fig. 9-18?

    A. The voltage at point x should be positive.

    B. The voltage at point x should be negative.

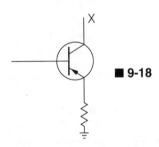

■ 9-18

26. The full-scale deflection of a certain meter movement is 0.01 milliamperes. What is the ohms-per-volt sensitivity rating of that meter movement?

    A. 0.01 ohms per volt

    B. 100 ohms per volt

    C. 20,000 ohms per volt

    D. None of these choices is correct.

27. Consider the circuit of Fig. 9-19 in which an ammeter is connected in series with a resistor. This connection will convert the ammeter to:

    A. a voltmeter.

    B. a wattmeter.

    C. an ohmmeter.

    D. None of these choices is correct.

9-19

28. In the circuit of Fig. 9-20 the resistance of R is to be measured. One way to do this is to use a meter that has a low-power ohms scale. That prevents the transistor from being turned on by the ohmmeter voltage. If a low-power ohms scale is not available, you could make this measurement by connecting the:

    A. negative side of the ohmmeter to the base side of the resistor and positive side of the ohmmeter to common.

    B. positive side of the ohmmeter to the base side of the resistor and negative side of the ohmmeter to common.

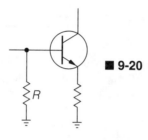

9-20

29. In Fig. 9-21, what is the frequency of the waveform being displayed?

    A. 50,000 hertz

    B. 100,000 hertz

    C. 150,000 hertz

    D. None of these choices is correct.

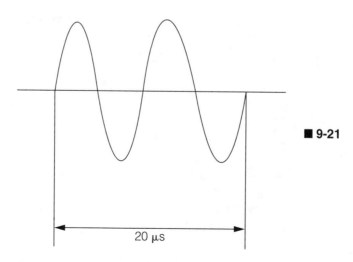

9-21

20 μs

30. What is the voltmeter reading for the ac circuit shown in Fig. 9-22?

   A. 5 volts
   B. slightly over 7 volts
   C. 10 volts
   D. The answer cannot be determined.

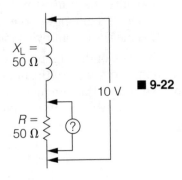

$X_L = 50\ \Omega$

$R = 50\ \Omega$

10 V   ■ **9-22**

31. The purpose of the circuit in Fig. 9-23 is to measure the leakage current of the capacitor. What type of meter should be used for this measurement?

   A. Ammeter
   B. Voltmeter

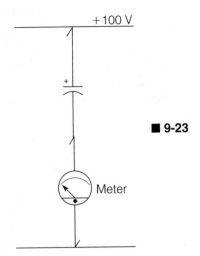

+100 V

■ **9-23**

Meter

32. Consider the circuit in Fig. 9-24. When the emitter is shorted to the base:

   A. V_1 will display 0 volts.
   B. V_1 will display the B+ voltage.

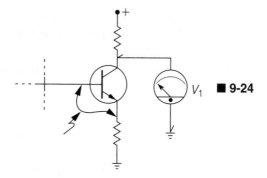

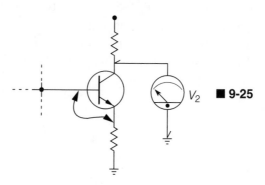

33. In the circuit of Fig. 9-25, when the emitter is shorted to the base:

    A. the reading of V_2 will be 0 V.
    B. V_2 will display the $B+$ voltage.

■ 9-24

■ 9-25

34. What will be the effect of shorting the collector to the base of a transistor as shown in Fig. 9-26?

    A. The transistor will be destroyed.
    B. No damage will occur.

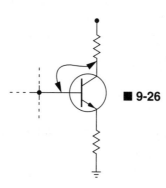

■ 9-26

35. Figure 9-27 shows a bipolar transistor with its base connected to its collector. Which of the following is true?

    A. The transistor will be destroyed by the connection.

    B. The transistor can be used as a diode when it is connected this way.

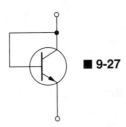

■ 9-27

36. A dc voltmeter is connected across a pure sine wave voltage source as shown in Fig. 9-28. The meter will display:

    A. the RMS value of the voltage.

    B. the peak value of the voltage.

    C. one-half the peak value of the wave.

    D. 0 V.

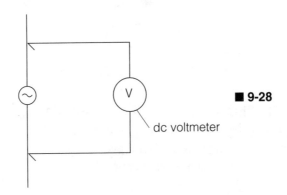

dc voltmeter

■ 9-28

37. The dc voltmeter in Fig. 9-29 should display:

    A. 6 V.

    B. 12 V.

    C. 13.07 V.

    D. about 0 V.

38. Which of the following is an example of a volatile memory?

    A. Static RAM

    B. Magnetic tape

    C. Floppy disk

    D. None of these choices is correct.

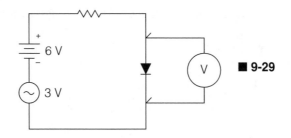

■ 9-29

39. Which equation best describes the gate in Fig. 9-30?

    A. $y = \overline{AB}$
    B. $y = \overline{A} + \overline{B}$
    C. $y = \overline{A} + \overline{B}$
    D. $y = A \times B$

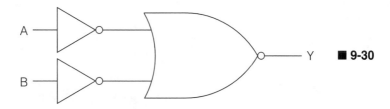

Y   **■ 9-30**

40. Convert $1AF_{16}$ to a decimal number.

    A. 431_{10}
    B. 263_{10}
    C. 315_{10}
    D. 511_{10}

41. Which of the following could be used for a parallel-to-serial conversion?

    A. Shift register
    B. Stack pointer
    C. Demultiplexer
    D. Both A and B

42. Which of the following is a symbolic statement of a DeMorgan's theorem?

    A. $\overline{\overline{A + B}} = \overline{A} + \overline{B}$
    B. $\overline{A}\ \overline{B} = \overline{AB}$
    C. $\overline{A + B} = \overline{A} \times \overline{B}$
    D. $\overline{A} + \overline{B} = \overline{AB}$

43. Identify the volatile memory in the following choices:

    A. ROM
    B. PROM
    C. EPROM
    D. None of these choices is correct.

44. Identify the circuit in Fig. 9-31:

    A. Full adder
    B. Half adder
    C. Flip-flop
    D. Comparator

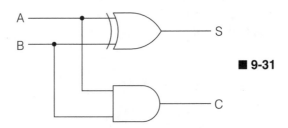

■ 9-31

45. Select the truth table for the gate in Fig. 9-32:

A.			B.			C.			D.		
A	B	Y	A	B	Y	A	B	Y	A	B	Y
0	0	0	0	0	1	0	0	1	0	0	1
0	1	1	0	1	1	0	1	0	0	1	0
1	0	1	1	0	0	1	0	0	1	0	1
1	1	0	1	1	1	1	1	1	1	1	1

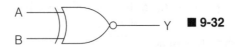

■ 9-32

46. The maximum load that a logic gate load can handle is known as:

    A. fan in.
    B. fan out.
    C. turnover.
    D. I^2L.

47. How is an EPROM erased in preparation for introducing a new program?

    A. With a strong magnetic field
    B. With ultraviolet light
    C. With visible light
    D. None of these choices is correct.

48. How many flip-flops are required to make 16 counts?

   A. 2
   B. 3
   C. 4
   D. 5

49. What is a parity bit used for?

   A. Error detection
   B. Addition
   C. Subtraction
   D. None of these choices is correct.

50. What is the forward voltage drop across an operating LED?

   A. 1.4 to 2.3 V
   B. 3 to 4 V
   C. 0.2 to 0.7 V
   D. 3.4 to 6 V

51. The expression $A + A$ reduces to:

   A. A.
   B. $\overline{A}$.
   C. 0.
   D. 1.

52. The circuit of Fig. 9-33 is equivalent to:

   A. $\overline{A}$ OR $\overline{B}$.
   B. $\overline{A}$ AND $\overline{B}$.
   C. an inverter.
   D. a NAND gate.

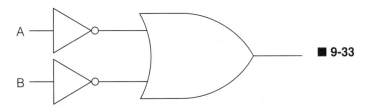

■ 9-33

53. Which of the following is an example of ROM operation?

   A. Storing tabular data
   B. Storing combinational logic functions
   C. Decoding
   D. All of these are correct.

54. A logic probe is basically:

    A. an ammeter.
    B. a voltmeter.
    C. a wattmeter.
    D. a signal injector.

55. A static RAM is constructed from:

    A. AND gates.
    B. flip-flops.
    C. both of the above.
    D. neither of the above.

56. A multiplexer could also be called a:

    A. data distributor.
    B. data selector.
    C. sequential-logic circuit.
    D. None of these choices is correct.

57. Which of the following is not correct?

    A. $A + A = A$
    B. $A + 0 = 0$
    C. $A \times A = A$
    D. $1 + A = A$

58. Another name for a *flip-flop* is:

    A. astable multivibrator.
    B. one shot.
    C. bistable multivibrator.
    D. monostable multivibrator.

59. Write an equation for Y in the circuit of Fig. 9-34.

    A. $(A + B) \, (\overline{C} + D)$
    B. $\overline{AB} + \overline{CD}$
    C. $(A + B) \times (C + D)$
    D. $AB + CD$

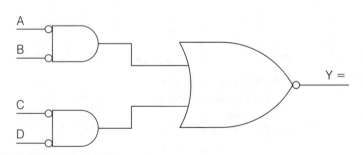

■ 9-34

60. A digital signal is:

    A. continuous.

    B. discrete.

    C. sinusoidal.

    D. None of these choices is correct.

61. The number of states a counter goes through before reset is called:

    A. the modulo.

    B. the increment.

    C. the decrement.

    D. None of these choices is correct.

62. A recirculating shift register is called:

    A. a buffer.

    B. a ring counter.

    C. a Schmitt trigger.

    D. an integrator.

63. The time it takes a circuit to respond to the leading edge of a pulse is called:

    A. rise time.

    B. delay time.

    C. fall time.

    D. storage time.

64. Which of the following must be refreshed?

    A. Static RAM

    B. Dynamic RAM

    C. ROM

    D. PROM

65. The circuit of Fig. 9-35 shows a simple logic probe. When the input signal to the probe is a logic 0:

    A. L1 is on, L2 is off.

    B. L1 is off, L2 is on.

    C. L1 and L2 are both off.

    D. L1 and L2 are both on.

66. An R-S flip-flop would not normally be used as a:

    A. data latch.

    B. bounceless switch.

    C. bistable latch.

    D. toggled flip-flop.

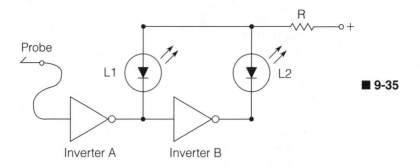

9-35

67. The following truth table is for:

    A. an exclusive NOR.
    B. an inclusive OR.
    C. an exclusive OR.
    D. None of these choices is correct.

Truth Table

A	B	C	L
0	0	0	0
0	0	1	0
0	1	0	0
0	1	1	0
1	0	0	0
1	0	1	0
1	1	0	0
1	1	1	1

68. Which of the following counters places less demand on the power supply?

    A. Synchronous
    B. Asynchronous

69. Which of the following counters is faster?

    A. Synchronous
    B. Asynchronous

70. A certain counter is made with five flip-flops. What is the highest number the counter could display?

    A. 15_{10}
    B. 24_{10}
    C. 31_{10}
    D. 32_{10}

71. In the circuit of Fig. 9-36, the switch is closed for 10 seconds and then opened. The lamp will:

    A. not turn on.

    B. turn on and stay on.

    C. turn on only for as long as the switch is closed.

    D. be on at all times regardless of the switch position.

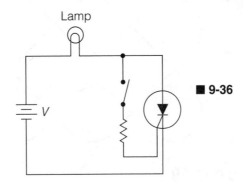

■ 9-36

72. The SCR in Fig. 9-36 is started into conduction:

    A. by a gate voltage.

    B. by a gate current.

73. What is the name of a permanently charged dielectric?

_____

74. What type of resistor is usually connected between terminals x and y in Fig. 9-37 to protect the transistor?

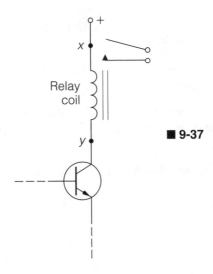

■ 9-37

75. Refer to the Lissajous figure in Fig. 9-38. Assume that it is displayed on an oscilloscope screen. What is the ratio of the horizontal frequency to the vertical frequency?

$$\frac{Horizontal\ frequency}{Vertical\ frequency} = ?$$

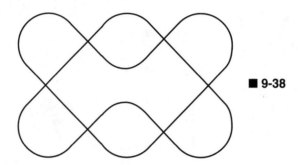

■ 9-38

Answers to quiz #9

1. The correct answer is B. The color codes most-often missed by technicians are: 10 ohms, 100 ohms, and for resistors having a resistance less than 1 ohm.

268

2.

$$P = \frac{V^2}{R} = \frac{20^2}{10} = 40 \text{ watts}$$

3. The correct answer is A. By ohm's law the voltage equals the current times the resistance:

$$V = I \times R$$
$$= 0.02 \text{ A} \times 2000 \text{ ohms}$$
$$V = 40 \text{ volts}$$

4. The correct answer is A. The charge path is shown by the arrow marked c.

5. The correct answer is A. When the temperature decreases the capacitance will increase. That, in turn, will result in a longer time constant.

6. The correct answer is D. The time constant equation for R-L circuits is:

$$\frac{T}{R} = L$$

where: L is in henries and R is in ohms

CET study guide

In this particular case:

$$T = \frac{L}{R} = \frac{0.250}{250} = 0.001 \text{ second} = 1 \text{ millisecond}$$

7. The correct answer is B. When the switch is in the y position, it is a step-up transformer with a higher secondary voltage.

8. The correct answer is A. You know this without having to work a problem because you know ohm's law. Working a lot of ohm's law problems gives you the insight needed for answering questions like this.

9. The correct answer is C. The voltages add as vectors like reactances and resistances. Figure 9-39 shows the graphical solution. Note that V_R and V_C are combined at right angles and the hypotenuse is the resultant voltage (V). A mathematical solution can also be obtained as follows:

$$V = \sqrt{V_R^2 + V_C^2} = \sqrt{6^2 + 10^2} = \sqrt{136} = 11.66 \text{ V}$$

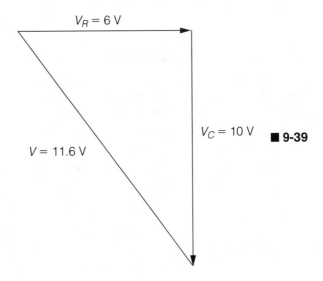

$V_R = 6$ V

$V_C = 10$ V ■ 9-39

$V = 11.6$ V

10. The correct answer is B. The circuit is redrawn in Fig. 9-40 to show the equivalent voltage source of 30 V and the common point.

First: Find the voltage across R1 by the proportional method.

$$V_1 = (V)\ \frac{R_1}{R_1 + R_2} = (30)\ \frac{10}{10 + 20} = 10 \text{ volts}$$

Second: Subtract the voltage across R1 from +24 V.

$$V_X = 24 - 10 = 14 \text{ V}$$

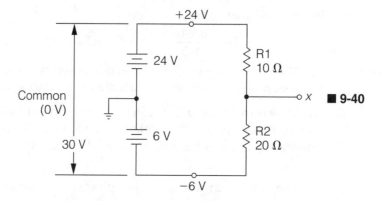

+24 V

24 V

Common
(0 V)

30 V

6 V

−6 V

R1
10 Ω

R2
20 Ω

○ X   ■ 9-40

11. The correct answer is A. According to the maximum power transfer theorem, the maximum amount of power will be delivered to a load resistor when its resistance is equal to the internal resistance of the source.

12. The total power in the circuit is:

$$Total\ power = \frac{V^2}{R} = \frac{24^2}{120} = 4.8\ watts$$

Half of that power is dissipated by R_L, so the answer is 2.4 watts.

13. The correct answer is A. The emitter of an NPN transistor must be negative with respect to the collector. Because the collector is at a common point, the emitter must be negative with respect to common.

14. The correct answer is D. The emitter current is: 3 V/30 Ω = 0.1 A = 100 mA. The base current is: 0.0018 − 0.0009 amperes.

Using the equation for currents in a transistor:

$$I_C = I_E - I_B = (0.1 - 0.0009)\ A = 0.0991 = 99.1\ mA$$

15. Each horizontal block represents 50 microseconds. The sine wave takes 8 horizontal blocks, so one complete cycle takes:

$$8 \times 50 = 400\ microseconds$$

Using the equation:

$$Frequency = 1/time\ for\ one\ cycle:$$

$$f = \frac{1}{T} = \frac{1}{400\ \mu s} = 2500\ hertz\ (answer)$$

16. Each vertical block represents 1.5 volts. Because there are 10 vertical blocks the peak-to-peak voltage is 15 volts. That means the peak voltage is 15/2 = 7.5 volts. Then:

$$V = 0.707 \times V_{MAX} = 0.707 \times 7.5 = 5.3\ V$$

Observe that no subscripts are used for RMS voltage.

270

17. The correct answer is B. Pushing the turns closer together increases the inductive coupling between the turns and increases the inductance. The resonant frequency of a series-tuned LC circuit is calculated by the equation:

$$f_r = \frac{1}{2\pi\sqrt{LC}}$$

Increasing the inductance (L) in the denominator reduces the value of the fraction, so the frequency is reduced.

18. The correct answer is A. In a PNP transistor, the collector is more negative than the base. That is another way of saying that the base is positive with respect to the collector.

19. The correct answer is B. You see this connection most often in high-voltage rectifier circuits.

20. The correct answer is D. However, you shouldn't use electrolytic capacitors in vacuum-tube RC-coupled circuits because leakage puts a positive voltage on the grid of the second amplifier. That positive voltage can destroy the second amplifier tube. The base of the second RC-coupled NPN transistor is already positive, so a small amount of leakage will not destroy the second amplifier. Because the voltage across the capacitor is usually low in transistor circuits and the electrolytic will pass low frequencies, you can understand why electrolytic coupling capacitors are used for RC coupling in audio stages.

21. The correct answer is B. Remember that the arrow on semiconductor symbols points away from the N region and toward a P region. In the illustration, the arrow is pointing away from the P-type material in the gate.

22. The correct answer is A. In both circuits, the zener diode is forward biased so it cannot work properly. Also, the forward-biased zener diode has only about 0.7 volts across it. That is practically a short circuit in a half-wave rectifier that operates directly off the line voltage.

23. The correct answer is B. The resistor is called a *surge-limiting resistor*. It usually has a value of less than 100 ohms. Its purpose is to limit the charging current for the electrolytic capacitors. That charging current flows through the diode. During the first few cycles of input voltage, that charging current could destroy the diode if the surge-limiting resistor was not in the circuit.

24. The correct answer is A. The circuit is an emitter follower. It is sometimes called a *common-collector circuit*. In this type of circuit, the output signal is in phase with the input circuit. The volt-

271

age gain of a follower circuit is less than 1.0, but the circuit can have a current gain greater than 1.0

25. The correct answer is B. For a PNP transistor, the collector should be negative with respect to the emitter.

26. The correct answer is D. The full-scale deflection is equal to the reciprocal of the ohms-per-volt sensitivity:

$$Ohms\ per\ volt = \frac{1}{Full\text{-}scale\ current} = \frac{1}{0.00001} = 100,000$$

27. The correct answer is A. The resistor is called a *series multiplier*.

28. The correct answer is A. By connecting the negative side of the ohmmeter to the base and the positive side to the emitter the ohmmeter, voltage will reverse bias the transistor emitter-base junction. That, in turn, eliminates the circuit in parallel with R and the ohmmeter will measure only the resistance of R.

29. The correct answer is B. The time for two cycles is 20 microseconds, so, the time for one cycle is 10 microseconds. The applicable equation $f = 1/T$ was given in a previous problem.

30. The correct answer is B. The inductive reactance and resistance are in quadrature. The applied 10 V is the resultant of the two voltages. Actually, 10 V is the hypotenuse of a right triangle with the two equal voltages as the legs:

$$V^2 + V^2 = 10^2$$
$$2V^2 = 100$$
$$V^2 = 50$$
$$V = 7.07\ V$$

31. The correct answer is B. If you connect an ammeter in the circuit it will likely be destroyed by the high charging current. If you know the resistance of the series multiplier resistor you can determine the leakage current.

32. The correct answer is B. The emitter-base short circuit removes the forward bias on the transistor. That shuts off the transistor (no collector current). Because there is no current flowing through the collector resistor there is no voltage drop across it and the voltmeter will display the full positive power supply voltage.

33. The correct answer is A. As in the previous question, the short circuit will shut off the transistor. No current will flow through the emitter resistor, so the voltage across it will be 0 V.

34. The correct answer is A. *Never* short the collector to the base in a transistor amplifier circuit.

35. The correct answer is B. This is not a transistor amplifier. The connection allows the transistor to be used as a diode.

36. The correct answer is D. The meter will display the full-cycle average of the voltage which is 0 V.

37. The correct answer is D. The diode is forward biased during the full cycle of ac. The voltage across the diode is less than 1 V.

38. The correct answer is A. A volatile memory is erased (over a short period of time) when the power supply voltage is removed.

39. The correct answer is B. The output can be written as $\overline{A} + \overline{B}$. Using DeMorgan's theorem (break the bar and change the sign) you get $\overline{\overline{A}} \times \overline{\overline{B}}$. The double overbars revert to $A \times B$.

40. The correct answer is A. Remember that you must be able to convert between numbers with different bases.

41. A

42. The correct answer is C. See Appendix E.

43. D

44. B

45. C

46. B

47. B

48. C

49. A

50. A

51. A

52. D

53. D

54. B

55. B

56. B

57. D

58. C

59. C

60. B

61. A

62. B

63. B

64. B

65. B

66. D

67. D (it is for a 3-input AND)

68. B

69. A

70. The correct answer is C. The counter can make 32 counts, but because the counts start with zero, the highest count you can make is 31.

71. B

72. B

73. Electret

74. VDR (voltage-dependent resistor)

75. 2/3

Practice for the associate-level CET test

CHAPTERS 1, 2, 3, 4, 5, AND 7 OF THIS BOOK ARE DIRECTLY related to the Associate CET test. The information in these chapters is also relevant to the Journeyman test. Remember that it is necessary to take the Associate test before you take the Journeyman test. Switches are normally shown in their open position on schematics—even though the question indicates that they are closed.

As with the Journeyman Practice Test in Appendix B, this test will show you the range of subjects covered in the actual test. It is not intended to be a question-for-question duplication of that test.

Try to find time to take each of the practice tests in this book in one sitting. That will help you to learn how to pace yourself. If you haven't taken a 75-question test lately, you might be surprised how much fatigue can set in before you finish. Being tired is a sure way to become careless and miss questions that you would normally answer correctly!

Instructions for Sections 1, 2, and 3

You will not be required to solve complicated math problems in this section. Solving j-operator problems is not required.

Be sure that you can determine the resistance, capacitance (capacity), and inductance of circuits that have series or parallel R, L, or C. You should know the color code for resistors. Be sure that you can calculate the allowable tolerance for resistors.

The voltages across reactive and resistive elements are in quadrature. In other words, you cannot add the voltage across a resistor and across a capacitor (or inductor) when they are in series.

Be sure that you can determine the frequency of a wave if you know the time for one cycle. The time-constant equations for RC circuits ($T = RC$) and RL circuits ($T = L/R$) are often required for this section.

Be able to find the time constants for RC and RL circuits.

Logic probes are used to determine dc levels in logic circuits. You might be asked to trace the circuit with a simple logic probe. Transistor, FET, and tube-operating voltages are often required for answering questions in this part of the test. Be able to trace currents in bridge circuits.

Remember that resistance in one or both legs of a parallel-tuned circuit does affect the resonant frequency of the circuit. The equations for X_L and X_C show you how these reactances are affected by changes in frequency or component values.

Basic mathematics

1. An oscilloscope displays two cycles of voltage waveform. To display three cycles:
   A. increase the sweep speed.
   B. decrease the sweep speed.
   C. adjust the vertical attenuator.
   D. None of these choices are correct.

2. Which of the following is the sum of 0011 and 0010?
   A. 2_5
   B. 2_8
   C. 2_{16}
   D. 5_{10}

3. What is the applied voltage (V) for the circuit in Fig. A-1?
   A. 16 volts
   B. 21 volts
   C. 30 volts
   D. 42 volts

4. What is the frequency of the square-wave signal in Fig. A-2?
   A. 5 kHz
   B. 0.05 MHz
   C. 80 kHz
   D. 0.05 MHz

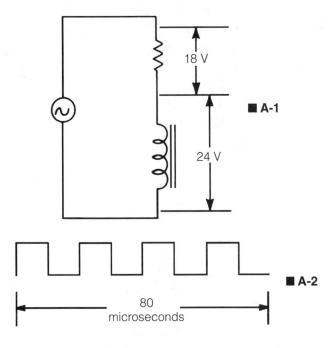

■ A-1

■ A-2

80
microseconds

5. Which of the following is true regarding Fig. A-3?
   A. If the tolerance of each resistor in A is +10%, the maximum resistance of the two resistors is 63.6 kΩ.
   B. The capacity of the combination in B is 10 μF.
   C. The time constant of the combination in C is 6 ms.
   D. The time constant of the combination in D is 0.5 μs.

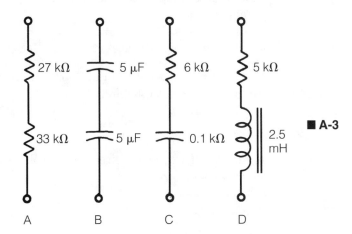

■ A-3

Dc circuits

6. The beta of the transistor in Fig. A-4 is 150. The base current should be:

    A. 500 microamperes.
    B. 50 microamperes.
    C. 5 microamperes.
    D. None of these choices is correct.

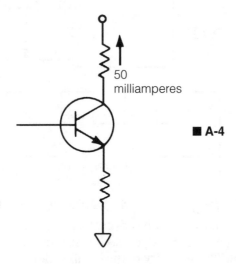

50 milliamperes

■ A-4

7. The emitter current for the transistor in question 6:

    A. cannot be determined from the information given.
    B. is 45 milliamperes.
    C. is 55 milliamperes.
    D. None of these choices is correct.

8. A simple logic probe is shown in Fig. A-5. In this case, LED #1 is glowing when the probe is touching:

    A. logic 1.
    B. logic 0.

9. Refer to Fig. A-6. The time required for the voltage across C to reach 63 volts is about:

    A. 2.7 microseconds.
    B. 27 milliseconds.
    C. 2.7 milliseconds.
    D. 0.27 milliseconds.

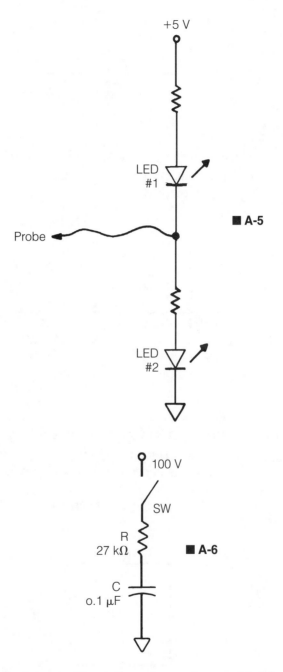

+5 V

LED
#1

Probe ◀

■ **A-5**

LED
#2

279

100 V

SW

R
27 kΩ   ■ **A-6**

C
o.1 μF

10. For the lamp in Fig. A-7 to glow at its rated brightness, the lamp current must be equal to:

    A. 1 amp.

    B. 0.4 amp.

    C. 333 milliamps.

    D. 0.2 amp.

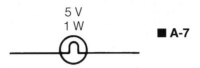

5 V
1 W

■ A-7

Ac circuits

11. Which of the waveforms in Fig. A-8 is leading?
   A. The one marked x.
   B. The one marked y.

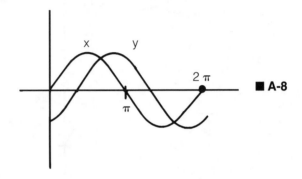

■ A-8

12. The purpose of R in the circuit of Fig. A-9 is to:
   A. lower the resonant frequency.
   B. increase the impedance of the circuit at resonance.
   C. broaden the frequency response.
   D. None of these choices is correct.

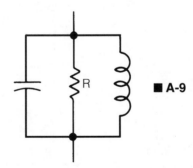

■ A-9

13. Which of the switch positions in Fig. A-10 will produce the highest secondary voltage?

    A. The one marked A.

    B. The one marked B.

    C. The one marked C.

    D. The answer cannot be determined from the information given.

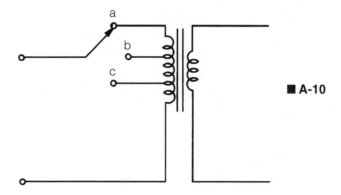

■ A-10

14. To increase the resonant frequency of a series LC circuit:

    A. move the capacitor plates closer together.

    B. move the capacitor plates farther apart.

15. The impedance of a series LC circuit is:

    A. maximum at resonance.

    B. minimum at resonance.

    C. not affected by frequency.

Transistors and semiconductors

Be sure that you know all of the dc voltages required for operating tubes, bipolar transistors, and field-effect transistors. Knowing the methods of obtaining bias for each type of amplifying device is important for answering some questions in this section. Be sure that you understand the operation of optoelectronic components (such as the optical coupler and seven-segment display).

16. The three leads for the components shown in Fig. A-11 are source, gate, and:

    A. collector.

    B. anode.

    C. plate.

    D. drain.

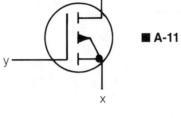

■ A-11

17. For the component in Fig. A-11 to work as an amplifier, the lead marked x should be:

    A. positive, with respect to the lead marked y.
    B. negative, with respect to the lead marked y.

18. Which of the following components would be connected between terminals x and y in Fig. A-12 to protect the transistor from inductive kickback?

    A. LDR
    B. VDR
    C. LED
    D. LAD

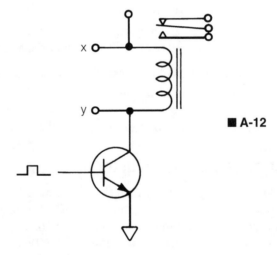

■ A-12

19. Switch SW in the circuit of Fig. A-13 is closed momentarily. When the switch is opened, the lamp:

    A. will stay on.
    B. will go off.

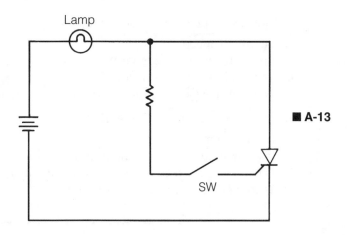

A-13

20. Which of the components in Fig. A-14 can be used as a very fast switch?

    A. The one marked A.
    B. The one marked B.
    C. The one marked C.
    D. The one marked D.

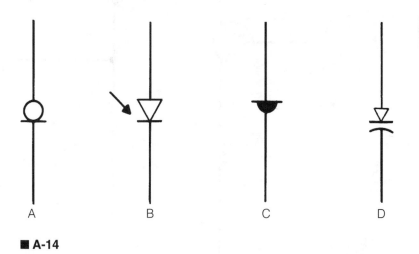

A-14

21. Which of the components in Fig. A-14 is normally operated with a reverse voltage?

    A. The one marked A.
    B. The one marked B.
    C. The one marked C.
    D. The one marked D.

22. An optical coupler has the advantage of:

    A. very high speed.
    B. operating with voltage but no current.
    C. very high isolation resistance.
    D. None of these choices are correct.

23. Regarding the circuit of Fig. A-15, Q2:

    A. will not amplify as shown.
    B. is connected as a common-emitter amplifier.
    C. is connected as a common-collector amplifier.
    D. is connected as a common-base amplifier.

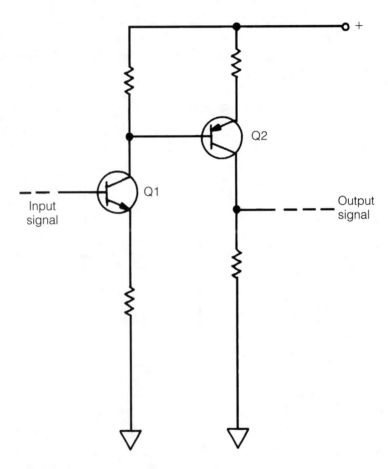

■ A-15

24. In the circuit of Fig. A-16, the output signal at A is:

    A. negative-going sawtooth.
    B. positive-going sawtooth.
    C. negative pulses.
    D. positive pulses.

25. To increase the output frequency in the circuit of Fig. A-16, move the arm of the variable resistor:

    A. toward x.
    B. toward y.

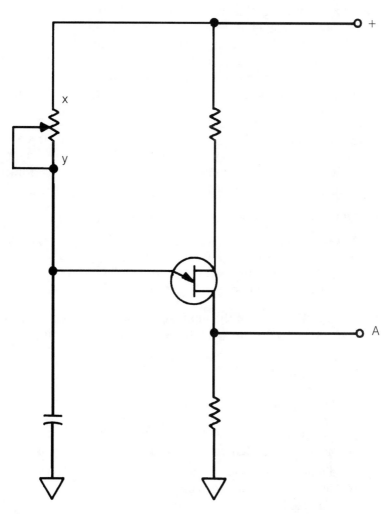

■ A-16

Electronic components and circuits

There is a certain amount of overlap between questions in some sections. After all, transistors and semiconductors are also electronic components and circuits. For that reason, you might find questions in this practice test that seem to be appropriate for a different section. This test makes the same distinctions that are used for making the actual test.

Remember that everyone who wants to become a Journeyman CET must take the Associate test. You should not expect to find specialized circuits in this test. All types of electronic systems use power supplies, amplifiers, oscillators, and operational amplifiers. So, you can expect to see questions on these subjects. Thyristors are appropriate study material for this section. Many of the circuits that use these components utilize time-constant circuits. You should not expect the circuits in this section to be drawn in a standard configuration.

When you are analyzing circuits, make it a practice to trace dc and/or ac signal paths to determine the purpose of components. That way you will not be confused by the way the circuit is drawn. A few questions in basic logic circuitry are typical for this section.

26. Which of the following describes the circuit in Fig. A-17?

    A. Colpits oscillator
    B. Long-tail biased amplifier
    C. High-gain amplifier with regenerative feedback
    D. Bootstrap amplifier

27. Refer to Fig. A-18. If the input signal at A has a peak-to-peak value of 100 millivolts, the peak-to-peak voltage at B should be:

    A. 130 millivolts.
    B. $100 \div 13,000$ millivolts.
    C. $100 \times 33/13$ millivolts.
    D. None of these choices is correct.

28. A component, such as the one shown in Fig. A-19, is located on the base-lead of a high-gain voltage amplifier. Its purpose is to:

    A. eliminate the Miller effect.
    B. eliminate parasitic oscillations.
    C. increase low-frequency response.
    D. increase high-frequency response.

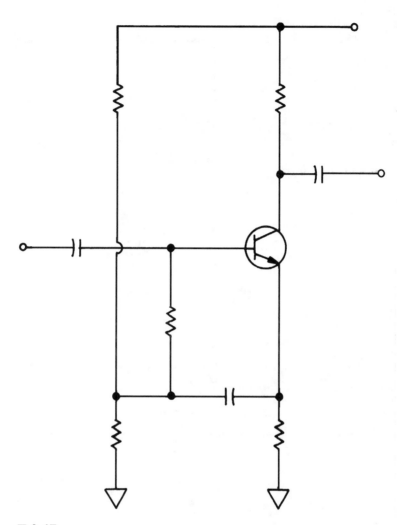

■ A-17

29. The component in the circuit of Fig. A-19 is called a:

    A. shunt capacitor.
    B. peaking bead.
    C. ferrite bead.
    D. bead ledge.

30. The symbol in Fig. A-20 represents:

    A. a VDR.
    B. a VOR.
    C. an IRE.
    D. a four-layer diode.

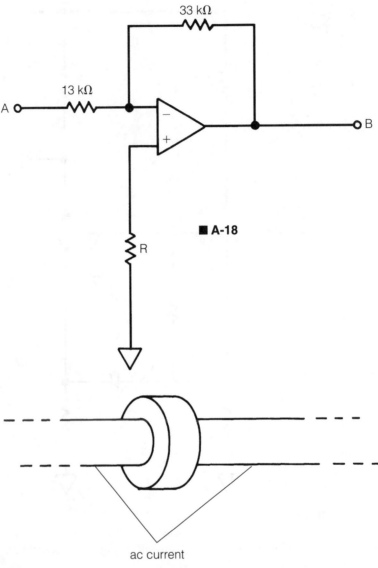

33 kΩ

13 kΩ

A

−

+

B

R

■ A-18

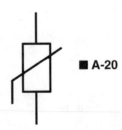

ac current

■ A-19

■ A-20

288

CET study guide

31. Solder used on printed circuit boards is often 60% tin and 40% lead. A better solder would be 63% tin and 37% lead because it is eutectic. In other words, it goes directly from a solid state to a liquid state when heated above the melting temperature. For surface-mount components a solder that melts at a lower temperature might contain as much as 5%:

    A. copper.
    B. silver.
    C. aluminum.
    D. gold.

32. Which of the following can be used to match a high impedance to a low impedance?

    A. Transformer
    B. Emitter-follower
    C. Pad
    D. All of these choices are correct.

33. Which of the following is a characteristic of an SCR?

    A. Once it starts conducting it cannot be shut off by a gate voltage input.
    B. It has a fast turn on when a pulse is delivered to the gate and the anode is positive, with respect to the cathode.
    C. It conducts only in one direction.
    D. All of these choices are correct.

34. Which of the following operates with reverse voltage, but not reverse current?

    A. Zener diode
    B. Diac
    C. Varactor diode
    D. Triac

35. A capacitor with a temperature coefficient rating of NPO:

    A. has a reduction of capacity as its temperature goes up.
    B. has an increase in capacity as its temperature goes up.
    C. increases in capacity as its temperature goes down.
    D. has a capacity that is not affected by temperature changes.

Instruments

You can expect questions on the oscilloscope and its uses. The emphasis is on test instruments used for troubleshooting. That includes add-on equipment, such as probes and prescalers. An understanding of both theory and use of test instruments is presumed for answering questions in this section. Some of the questions concern applications of test equipment.

36. Which of the following is most important for measuring the voltage drop across a 4.7-MΩ resistor?

    A. Taut-band meter movement
    B. Meter movement with a mirrored scale
    C. Meter with a high impedance
    D. ×10 meter probe

37. A VOM is used to measure the ac voltage of Fig. A-21. The meter should indicate a voltage value of:

    A. 10 volts.
    B. 7.07 volts.
    C. 6.36 volts.
    D. None of these choices are correct.

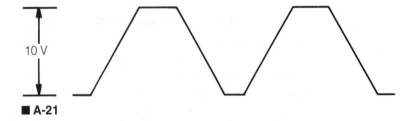

10 V

■ A-21

38. To convert a sensitive meter movement into an ac voltmeter you would need:

    A. a shunt and a rectifier.
    B. a shunt, but not a rectifier.
    C. a rectifier and a multiplier.
    D. None of these choices are correct.

39. Wheatstone bridge is usually used for accurate measurement of:

    A. current.
    B. voltage.
    C. resistance.
    D. power.

40. The oscilloscope display in Fig. A-22 shows three cycles of a sine wave. The frequency of the voltage being displayed is:

    A. 33.3 Hz.
    B. 200 Hz.
    C. 300 Hz.
    D. 333 Hz.
    E. None of these choices are correct.

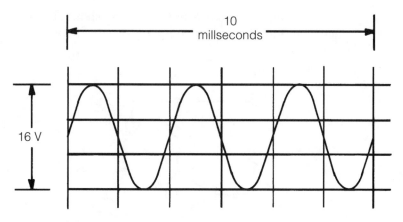

10
millseconds

16 V

■ **A-22**

41. The half-cycle average of the waveform in Fig. A-22 is about:
  A. 11.3 volts.
  B. 10.2 volts.
  C. 5.66 volts.
  D. 5.09 volts.

42. Assume that the oscilloscope display in Fig. A-22 is obtained by using a ×10 probe. The actual peak-to-peak voltage is:
  A. 1.6 volts.
  B. 160 volts.

43. The probe of Fig. A-23 has two LEDs—one to indicate logic 1 and one to indicate logic 0. A square-wave clock pulse would cause both LEDs to be:
  A. on.
  B. off.

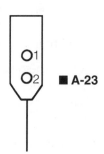

■ **A-23**

44. To make the voltage measurement shown in Fig. A-24, the negative side of the voltmeter goes to:

    A. point A.
    B. point B.

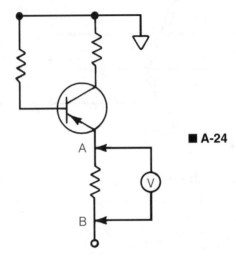

■ A-24

45. In a jeweled meter movement the jewels are:

    A. bearings.
    B. rubies.
    C. diamonds.
    D. sapphires.

Tests and measurements

Most of the questions in this section are about measurement of frequency, voltage, current, or resistance. You should understand the parameters for evaluating bipolar transistors and FETs. You are likely to be asked to interpret a Lissajous pattern.

Always remember that the name of each section refers to the emphasis for that section. Not every question on this subject is in this section. In some cases, a question might cut across two or three subjects, so it might be in any of three sections.

46. Which of the following is true regarding the amplifier bode plot in Fig. A-25?

    A. This curve cannot be obtained on an oscilloscope.
    B. This curve can be obtained on an oscilloscope.
    C. This curve is an example of a time domain display.

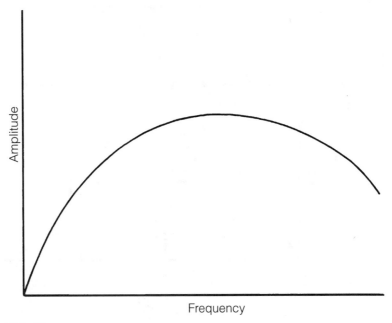

Amplitude

Frequency

■ A-25

47. The Z-axis on an oscilloscope is used to control:
   A. phasor displays.
   B. astigmatism.
   C. distortion.
   D. intensity.

48. A Lissajous test setup is shown in Fig. A-26. Assuming this is a class-A audio voltage amplifier, the display should be:
   A. a circle.
   B. an ellipse.
   C. a straight line.
   D. a sine wave.

49. You would not use the test setup of Fig. A-26 to:
   A. determine if clipping is present.
   B. measure intermodulation distortion.
   C. measure phase shift distortion.
   D. check for nonlinear amplification.

50. Figure A-27 shows the result of a square-wave test on a low-frequency class-A amplifier. The amplifier has a:
   A. poor low-frequency response.
   B. poor high-frequency response.
   C. dynamic overload.
   D. problem with clipping.

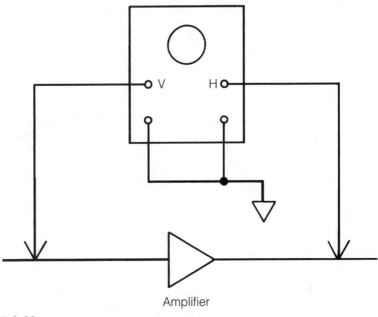

Amplifier

■ A-26

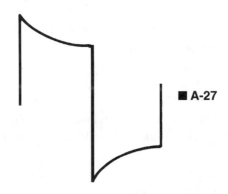

■ A-27

51. The bandwidth of an amplifier is determined on a bode plot. It is the range of frequencies between points where the output voltage is:

    A. between 10% and 90% of maximum.
    B. between the points where the output is down to 90% of maximum.
    C. between the points where the output is down to 70.9% of maximum.
    D. between the points where the output is down to 50% of maximum.

52. Consider the Lissajous pattern in Fig. A-28. It is obtained by comparing two sine-wave voltages that have different frequencies. If the vertical frequency is 1200 Hz, the horizontal frequency must be:

    A. 1600 Hz.
    B. 1200 Hz.
    C. 1150 Hz.
    D. 900 Hz.

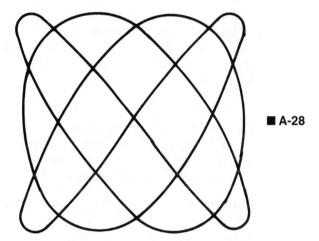

■ A-28

53. The ripple frequency of a full-wave rectifier should be:

    A. 60 Hz.
    B. 120 Hz.

54. Two different pure sine-wave frequencies are delivered to a linear amplifier at the same time. An oscilloscope display shows that more than two frequencies are at the output. The problem is:

    A. ringing.
    B. antiresonance.
    C. intermodulation distortion.
    D. I/f noise gain.

55. An advantage of sawtooth testing over square-wave testing is that the sawtooth test:

    A. is easier to read.
    B. shows clipping.
    C. is easier to set up.
    D. costs less.

Troubleshooting and circuit analysis

In this section, you might find questions about square-wave, triangular wave, and tone-burst tests. You can expect questions on the phase relationships between the input and output signals of an amplifier.

Be sure that you know the difference between signal injection and signal tracing. With signal injection you supply the signal to each section. An output signal indicates that everything is ok between the point where the signal is injected and the output section. With signal tracing, you follow the signal from the input to the output. When you lose that signal, you have just passed the source of the trouble. Disregard questions that deal with TV. They are for the Journeyman test.

56. You would expect to find a snubber in:

    A. a triac circuit with a resistive load.
    B. a power supply ferroresonant transformer circuit.
    C. an SCR circuit with an inductive load.
    D. direct-coupled bipolar transistor amplifiers.

57. Study the input signals to the logic gate in Fig. A-29. The output should be:

    A. a logic 0 at all times.
    B. a logic 1 at all times.

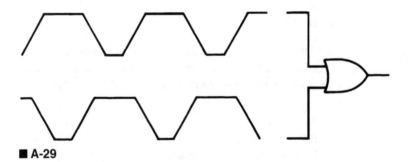

■ A-29

58. The audio amplifier in Fig. A-30 is:

    A. cut off.
    B. saturated.

59. Closing switch SW in the low-frequency amplifier circuit of Fig. A-30 will:

    A. broaden the frequency response of the amplifier.
    B. increase the signal gain of the amplifier.

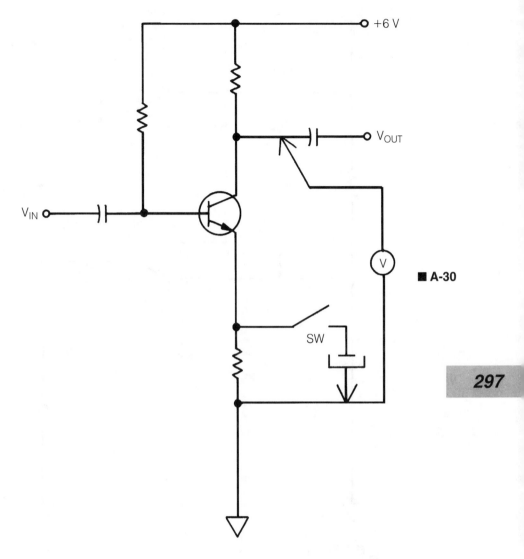

+6 V

V_{OUT}

V_{IN}

V

■ A-30

SW

60. Refer to Fig. A-31. Disregarding the forward voltage drop across the diodes, the output for the switch positions shown is:

    A. logic 1.
    B. logic 0.

61. Parallel-line transmission line is usually matched to coaxial cable by using:

    A. an emitter-follower.
    B. a common-base amplifier.
    C. a quadrac.
    D. a balun.

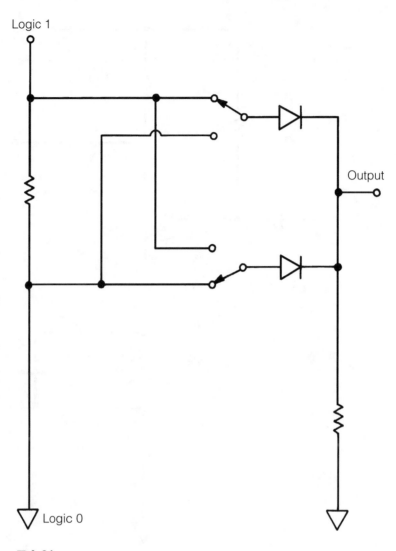

Logic 1

Output

Logic 0

■ A-31

62. Which of the following is true about a hot-carrier diode?

    A. It has a very low forward-voltage drop.
    B. It has a very low reverse voltage rating.
    C. It is used to heat crystals in a crystal oven.
    D. It is useful only at extremely high frequencies.

63. Refer to Fig. A-32 resistor R1:

    A. is used to limit load current.
    B. is never used.
    C. is a surge-limiting resistor.
    D. should be 10 kΩ or higher.

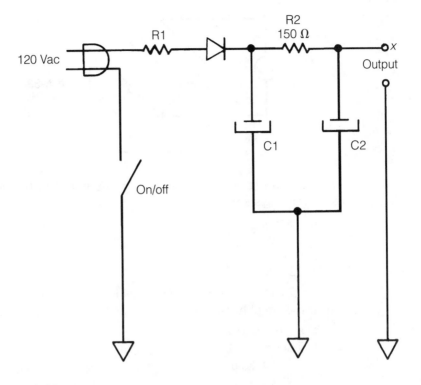

120 Vac

R1

R2
150 Ω

C1   C2

On/off

Output

x

■ A-32

64. Refer to Fig. A-32. With a 10-kΩ resistor across the output terminals you would expect the voltage at point X to be about:

    A. 90 volts.

    B. 119 volts.

    C. 165 volts.

    D. None of these choices is correct.

65. A 100-foot 300-ohm transmission line is cut into two 50-foot lengths. The impedance of each half is:

    A. 150 ohms.

    B. 225 ohms.

    C. 300 ohms.

66. The circuit of Fig. A-33:

    A. is a bridge rectifier.

    B. is not a bridge rectifier.

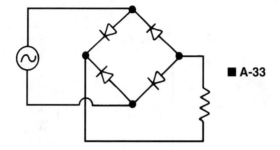

■ A-33

67. Refer to Fig. A-34. If this is a class-A amplifier, the dc volt-meter reading should be about:

    A. 3 volts.

    B. 6 volts.

    C. 9 volts.

    D. 12 volts.

68. If a 1000-Hz pure sine-wave signal is delivered to the input of the class-A amplifier in Fig. A-34, the dc voltmeter should show:

    A. an increase in voltage.

    B. a decrease in voltage.

    C. no change in voltage.

    D. 0 volts.

69. The amplifier in Fig. A-35 is operating:

    A. class A.

    B. class B.

    C. class C.

    D. as a cathode follower.

70. Is the following statement correct regarding the circuit of Fig. A-35? The grid current should be equal to about 5% of the cathode current.

    A. Correct

    B. Not correct

71. Is the following statement correct regarding the amplifier in Fig. A-36? The supply voltage (V) could be 200 volts.

    A. Correct

    B. Not correct

72. If y is shorted to x in the circuit of Fig. A-36:

    A. the device will be destroyed.

    B. the voltage at z will be equal to V to almost zero volts.

    C. R1 will act as a fuse and burn out thereby protecting the amplifying device.

    D. None of these choices is correct.

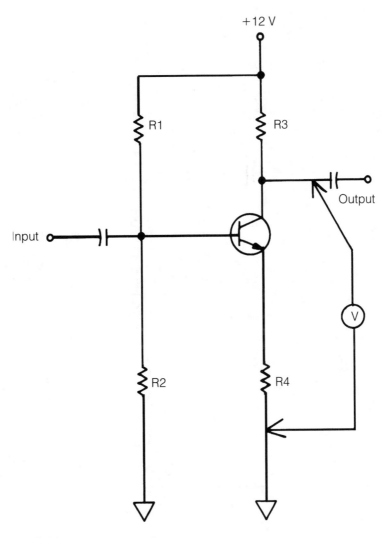

+12 V

R1

R3

Input

Output

V

R2

R4

A-34

73. For the class-A amplifier of Fig. A-36:
    A. y should be negative with respect to x.
    B. v should be negative.
    C. z should be negative with respect to x.
    D. All of the choices are correct.

74. In the amplifier circuit of Fig. A-37:
    A. no current can go through R2 unless there is an input signal.
    B. bias is established by the voltage drop across R1.
    C. bias is established by the voltage drop across R3.
    D. voltage V should be negative.

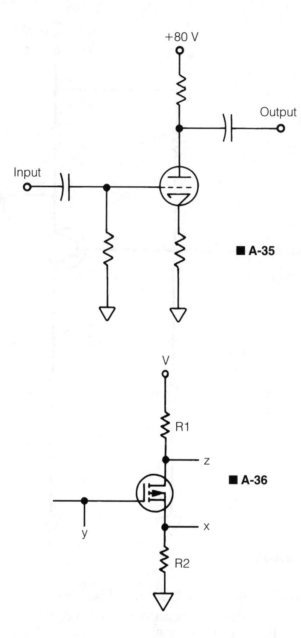

+80 V

Output

Input

■ A-35

V

R1

z

■ A-36

y

x

R2

75. Each resistor in the circuit of Fig. A-38 is 100 ohms. Resistor R3 is open. The voltmeter should indicate a voltage of about:

A. 0 volts.
B. 6 volts.
C. 12 volts.
D. None of these choices is correct.

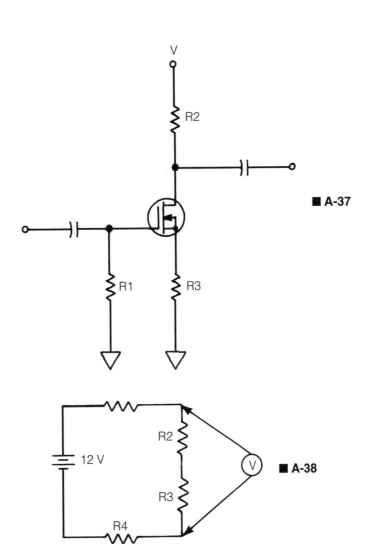

■ A-37

■ A-38

Answers to the practice test

Question	Answer
1.	B
2.	D
3.	C
4.	B
5.	D
6.	D
7.	D
8.	B
9.	C

Question	Answer
10.	D
11.	A
12.	C
13.	C
14.	B
15.	B
16.	D
17.	A
18.	B
19.	A
20.	C
21.	D
22.	C
23.	B
24.	D
25.	A
26.	D
27.	C
28.	B
29.	C
30.	A
31.	B
32.	D
33.	D
34.	C
35.	D
36.	C
37.	D
38.	C
39.	C
40.	C
41.	D
42.	B
43.	A
44.	A
45.	A
46.	B
47.	D
48.	C
49.	B
50.	A
51.	C
52.	D

Question	Answer
53.	B
54.	C
55.	B
56.	C
57.	B
58.	A
59.	B
60.	A
61.	D
62.	A
63.	C
64.	C
65.	C
66.	A
67.	B
68.	C
69.	A
70.	B
71.	A
72.	B
73.	D
74.	C
75.	C

Practice test for consumer products

B

76. Figure B-1 shows the symbol for:
   A. an OR gate.
   B. a NOR gate.
   C. an exclusive OR gate.
   D. an AND gate.
   E. an inverter.

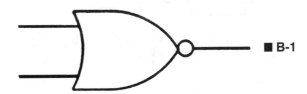

■ B-1

77. The arrow in Fig. B-2 points to:
   A. pin 7.
   B. pin 12.

■ B-2

78. Is this statement correct? *If you measure the logic levels (shown on the gate in Fig. B-3) it must be working properly.*
   A. The statement is correct.
   B. The statement is not correct.

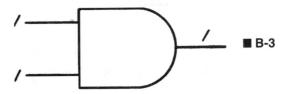

■ B-3

79. Which of the following might be used to delay a signal?

A. Flip-flop
B. Diplexer
C. Duplexer
D. Multiplexer
E. Bucket brigade

80. For the input signals shown to the gate in Fig. B-4, the output should be a:

A. pulse.
B. square wave.
C. logic 0 at all times.
D. logic 1 at all times.
E. sawtooth waveform.

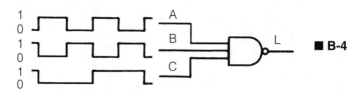

81. The input marked C on the gate in Fig. B-5 is permanently connected to common. Now the output will be a:

A. pulse.
B. square wave.
C. logic 0 at all times.
D. logic 1 at all times.
E. sawtooth waveform.

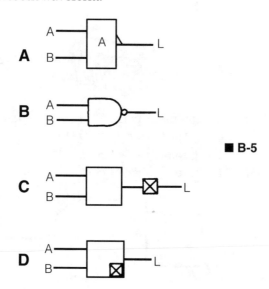

82. Which of the symbols in Fig. B-5 represents a NAND gate?

    A. Only the one marked A
    B. Only the one marked B
    C. Only the one marked C
    D. Only the one marked D
    E. All are symbols used to represent NAND gates.

83. Figure B-6 shows a relay logic circuit for:

    A. a NAND.
    B. an OR.
    C. an exclusive OR.
    D. a NOR.
    E. an AND.

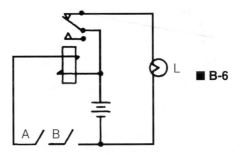

■ B-6

84. The truth table in Fig. B-7 is for:

    A. a NAND.
    B. an OR.
    C. an EXCLUSIVE OR.
    D. a NOR.
    E. an AND.

A	B	C	L
0	0	0	0
0	0	1	1
0	1	0	1
0	1	1	1
1	0	0	1
1	0	1	1
1	1	0	1
1	1	1	1

■ B-7

85. Is this equation correct: $A + B = AB$?

    A. It is correct.
    B. It is not correct.

86. Which of the following logic families is normally the fastest?

    A. TTL
    B. CMOS
    C. ECL
    D. RTL
    E. DTL

87. Which of the following logic families operates with a –5-volt supply?

    A. TTL
    B. CMOS
    C. ECL
    D. RTL
    E. DTL

88. For the logic probe in Fig. B-8, the LED marked X will be on when the probe tip touches:

    A. logic 0.
    B. logic 1.

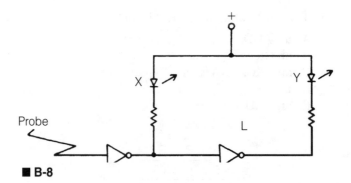

■ B-8

89. To make a RAM the manufacturer might use:

    A. fuses.
    B. flip-flops.
    C. decoders.
    D. encoders.
    E. None of these choices is correct.

90. What is the decimal equivalent of binary 1101?

    A. 10

    B. 11

    C. 12

    D. 13

    E. 14

Section 11: Linear circuits in consumer products

91. Which of the following amplifier configurations is best for high-frequency operation?

    A. Common emitter

    B. Common base

    C. Common collector

92. Which of the following is the same as beta?

    A. h_{FB}

    B. h_{FC}

    C. h_{FE}

    D. h_{FR}

    E. None of these choices is correct.

93. Whenever you increase the gain of an amplifier, you automatically:

    A. increase the bandwidth.

    B. decrease the bandwidth.

94. In the phase-locked loop in Fig. B-9, the block marked X is:

    A. a flip-flop.

    B. a counter.

    C. a decoder.

    D. an operational amplifier.

    E. a VCO.

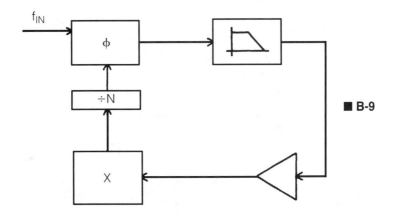

■ B-9

95. In the operational amplifier circuit in Fig. B-10, the maximum gain occurs when switch SW is in position:

    A. X
    B. Y
    C. Z

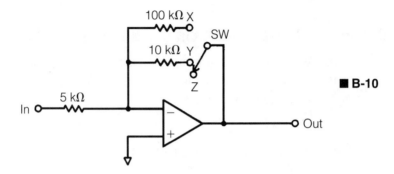

■ B-10

96. The input stage to an integrated circuit operational amplifier is:

    A. a totem pole.
    B. a two-input NOR.
    C. a differential amplifier.
    D. a Darlington pair.
    E. an R-S flip-flop.

97. To connect an operational amplifier in a common-mode configuration:

    A. ground the output.
    B. connect the output to the inverting input terminal.
    C. connect the output to the noninverting input terminal.
    D. connect the inverting and noninverting terminals together.
    E. connect the two input terminals and the output terminal to B+.

98. The circuit in Fig. B-11 is:

    A. useless.
    B. surely going to be destroyed.
    C. an oscillator.
    D. a buffer.

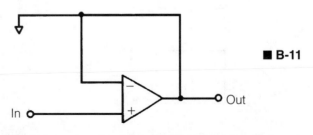

■ B-11

312

99. The battery in the parallel-tuned circuit of Fig. B-12 is:

    A. unnecessary.
    B. connected backwaards.
    C. properly connected.

100. Moving the arm of the variable resistor toward A in the circuit of Fig. B-12 will:

    A. not affect circuit resonance because the battery is connected backwards.
    B. increase the resonant frequency of the circuit.
    C. decrease the resonant frequency of the circuit.

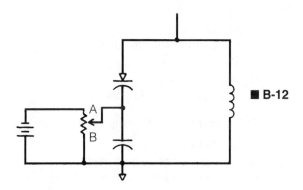

■ B-12

101. To decrease the high-frequency response of the circuit in Fig. B-13, you should move the arm of the variable resistor toward:

    A. point X.
    B. point Y.

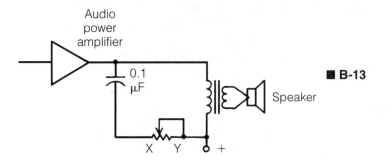

Audio power amplifier

0.1 μF

Speaker

■ B-13

102. You should expect to see an operational amplifier listed under:

    A. analog integrated circuits.
    B. thick film circuits.
    C. thin film circuits.

D. digital integrated circuits.

E. printed circuits.

103. Which of the following is most likely to produce level shifting?

A. NPN/PNP complementary direct-coupled pair

B. RC-coupled amplifier

C. Direct-coupled NPN transistors

D. NPN/PNP totem-pole configuration with long-tail bias

E. Transformer-coupled audio amplifiers

104. Which of the following is a method of rating differential amplifiers?

A. Beta squared

B. Common-mode rejection ratio

C. Capture ratio

D. Roll off

E. Intrinsic standoff ratio

105. A disadvantage of Darlington amplifiers in power circuits is:

A. low beta.

B. poor transient response.

C. poor low-frequency response.

D. high characteristic noise.

E. relatively high internal current.

106. A constant-current, two-terminal component can be made with a:

A. JFET and resistor.

B. metal-film capacitor.

C. self-saturating step-up transformer.

D. zener diode.

E. None of these choices is correct.

Television

107. Which of the following is a circuit that is used to keep the color oscillator on frequency?

A. AFC

B. AFT

C. AGC

D. AFPC

E. ALU

108. Some television receivers automatically adjust the tint using the:

   A. VITS signal.
   B. VIRS signal.

109. Which of the following is used to regulate the high voltage in a color receiver?

   A. VDR
   B. Zener diode
   C. Thermistor
   D. Negative feedback of the horizontal output stage
   E. Combination VDR and thermistor circuit

110. Colored snow during a black-and-white program might indicate:

   A. a defective low-voltage regulator.
   B. the ABL circuit is not working properly.
   C. the color killer isn't working properly.
   D. a defective antenna.
   E. None of these choices is correct.

111. Which of the following is likely to be used as a video detector in a television receiver?

   A. Slope detector
   B. Product detector
   C. Envelope detector
   D. Taylor demodulator
   E. None of these answers is correct.

112. To operate the keyed AGC circuit, it is necessary to have a composite video input and:

   A. a horizontal pulse from the AFC circuit.
   B. a pulse from the flyback.
   C. an AGC input.
   D. a signal from the AFT circuit.
   E. an input from the ABL.

113. You would expect to find a comb filter:

   A. in the audio section.
   B. used to separate video and color signals.
   C. in the tuner.
   D. in the vertical oscillator circuit.
   E. in the low-voltage power supply.

114. Peaking compensation would most likely be found in:

    A. the tuner.
    B. the power supply
    C. in the Y amplifier.
    D. in the sound section.
    E. in the high-voltage section.

115. A crowbar is:

    A. an overvoltage protection circuit.
    B. used to separate sound and picture signals.
    C. a burglar tool, not an electronic circuit.
    D. used to separate H and V signals.
    E. None of these choices is correct.

116. To synchronize a relaxation oscillator, the free-running frequency of the oscillator should be:

    A. higher than the sync signal.
    B. lower than the sync signal.

117. The brightness control is most likely to be located in the:

    A. high-voltage rectifier.
    B. horizontal output stage.
    C. horizontal oscillator.
    D. direct-coupled video amplifier stage.
    E. low-voltage power supply.

118. The sound takeoff point in a color television receiver is:

    A. before the video detector.
    B. at the video detector.
    C. after the video detector.

119. The sound IF frequency in a television receiver is:

    A. 41.75 MHz.
    B. 45.25 MHz.
    C. 27.5 MHz.
    D. 4.5 MHz.
    E. 120,000 Hz.

120. A phase-locked loop might be used in the:

    A. television tuner.
    B. low-voltage regulator.
    C. high-voltage regulator.
    D. ACL circuit.
    E. AGC circuit.

121. When you are facing the television receiver, the picture is scanned:

    A. from left to right.

    B. from right to left.

Videocassette and video disks

122. Which of the following is a disadvantage of a video-disc system?

    A. Higher cost than video tape systems

    B. Cannot be used to record

123. Which of the following is used to read signal information from a video disc?

    A. Sonar

    B. Radar

    C. Maser

    D. Laser

    E. Elliptical stylus

124. In a video tape recorder, the video signal is recorded as:

    A. AM.

    B. FM.

125. To maintain a constant tape speed, a VTR system uses:

    A. a cap screw.

    B. a governor on the motor.

    C. an accelerometer.

    D. Any of these might be used.

    E. None of these answers is correct.

126. Which of the following is the most widely used home TV recording system?

    A. VHS

    B. Cartridgevision

    C. Betamax

Troubleshooting consumer equipment

127. The waveform in Fig. B-14 has been obtained with a square-wave test. It shows a:

    A. loss of low frequencies.

    B. loss of high frequencies.

■ B-14

128. To check Q1 in the circuit of Fig. B-15, a technician shorts its emitter to its base and measures the collector voltage. This test:

A. is the best method of determining the condition of Q1.
B. will destroy Q1.
C. will destroy Q2.

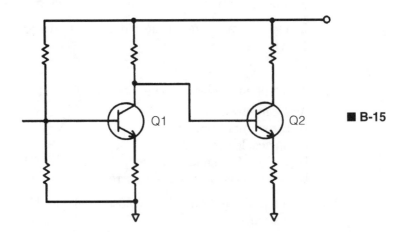

■ B-15

129. You would expect to find horizontal synchronizing pulses at the output of the:

A. differentiator.
B. integrator.

130. In a color television receiver, you would expect the color circuitry to be off when:

A. the color killer is on.
B. the color killer is off.

131. The time for one line of sweep in a television receiver is about:
   A. 123 microseconds.
   B. 94.3 microseconds.
   C. 59.8 milliseconds.
   D. 56.5 milliseconds.
   E. None of these choices is correct.

132. In which of the following sections would you expect to find the automatic degaussing circuit?
   A. Video amplifier
   B. High-voltage section
   C. Low-voltage section
   D. Horizontal sweep section
   E. Tuner

133. You might find an LDR in the:
   A. vertical sweep circuit.
   B. horizontal sweep circuit.
   C. low-drain resistor circuit.
   D. brightness circuit.
   E. high-voltage supply.

134. The ACC circuit is not working properly. This will cause improper gain in the:
   A. bandpass amplifier.
   B. color-sync circuit.
   C. video amplifier.
   D. tuner.
   E. None of these choices is correct.

135. The VITS signal is located on:
   A. the horizontal blanking pedestal.
   B. the vertical blanking pedestal.

136. The AGC voltage in a transistorized television receiver:
   A. should be a negative dc.
   B. should be a positive dc.
   C. could be either a positive or negative dc.
   D. might be a square wave.
   E. should be a sinusoidal ac.

137. Which of the following requires a startup circuit?
   A. Scan-derived supply
   B. Analog regulator
   C. High-voltage regulator

D. Phase-locked oscillator

E. AFT

138. If a surge-limiting resistor is burned out:

    A. replace it with one that has a higher power rating; never replace it with one that has a lower power rating.

    B. expect the high-voltage measurement to be high.

    C. no dc output will come from the low-voltage supply.

    D. the set must have been struck by lightning.

    E. it will destroy the audio power transistor.

139. Use a ringing test to check:

    A. continuity in the high-voltage section.

    B. the yoke.

    C. the video amplifiers.

    D. start the horizontal oscillator.

    E. AGC recovery time.

140. The IF frequency of a broadcast FM receiver is:

    A. 4.5 MHz.

    B. 10.7 MHz.

    C. 455 kHz.

    D. 41.25 kHz.

    E. None of these choices is correct.

141. When using a Lissajous pattern to check an audio amplifier for distortion, you should get:

    A. an ellipse.

    B. a circle.

    C. a square.

    D. a straight line.

    E. a pattern of dots.

Test equipment

142. Which of the following is a method of evaluating transistors in a common base configuration?

    A. Alpha

    B. Beta

    C. Gamma

    D. Slewing rate

    E. CMRR

143. Which of the following is used to convert a sensitive meter movement to a voltmeter?

    A. Shunt

    B. Multiplier

144. A meter movement with a full-scale deflection reading of 10 microamperes is used as a voltmeter. The meter has a rating of:

    A. 10,000 ohms per volt.

    B. 20,000 ohms per volt.

    C. 50,000 ohms per volt.

    D. 100,000 ohms per volt.

    E. The answer cannot be determined from the information given.

145. Which of the following would be useful for looking at the VITS signal?

    A. 5-MHz recurrent-sweep oscilloscope

    B. Prescaler

    C. Logic analyzer

    D. Vector scope

    E. Scope with a delayed sweep

Answers to practice test

Question	Answer
76.	B
77.	A
78.	B (There are other possibilities.)
79.	E
80.	D
81.	D
82.	E
83.	A
84.	B
85.	A
86.	C
87.	C
88.	B
89.	B
90.	D
91.	B
92.	C

Question	Answer
93.	B
94.	E
95.	A
96.	C
97.	D
98.	D
99.	B
100.	B
101.	A
102.	A
103.	C
104.	B
105.	E
106.	A
107.	D
108.	B
109.	D
110.	C
111.	C
112.	B
113.	B
114.	C
115.	A
116.	B
117.	D
118.	A
119.	D
120.	A
121.	A
122.	B
123.	D
124.	B
125.	E
126.	A
127.	B
128.	C
129.	A
130.	A
131.	E
132.	C
133.	D
134.	A
135.	B

Question	Answer
136.	C
137.	A
138.	C
139.	B
140.	B
141.	D
142.	A
143.	B
144.	D
145.	E

Computer practice test

THE AUTHORS STRONGLY ADVISE YOU TO OBTAIN STUDY guides for the computer option from ISCET before taking any CET test.

1. Binary numbers and other characters are displayed according to the ASCII (As-Key) code (American Standard Code for Information Interchange). Seven bits are used in combinations of ones and zeros. The total number of combinations is:

    A. $7^2 = 49$.
    B. $2^7 = 128$.

2. A storage section is in the shape of a wedge—it divides tracks on a floppy disk, thus making it easy to access specific data:

    A. Wedge.
    B. Pi access.
    C. Sector.
    D. Data plus.

3. A narrow section on a tape or disk where data is stored:

    A. Section.
    B. Track.
    C. D-line.
    D. Data path.

4. A combination of data bits treated together by a computer; it is capable of being stored in one or two address locations in memory:

    A. Track unit.
    B. Bit group.
    C. A-code.
    D. Word.

5. A type of disk that can accept data only one time and played back a number of times. It cannot be erased.

    A. Unitary disk
    B. Prime disk
    C. WORM
    D. Write right

6. Which of the following is not normally fed to a computer?

   A. Digits 0 through 9
   B. Bar codes
   C. Keyboard output
   D. Punched card readout

7. Step-by-step instructions on a computer are called:

   A. conversion bits.
   B. ready words.
   C. programs.
   D. None of these choices is correct.

8. Keyboards are usually connected to the computer through the:

   A. input terminal.
   B. output port.
   C. I/O port.
   D. lockup terminal.

9. The data bus connected to a ROM is a:

   A. two-way bus.
   B. one-way bus.

10. The data bus connected to a RAM is a:
    A. two-way bus.
    B. one-way bus.

11. Which of the following describes a tristate device?

    A. Passes data in either of two directions
    B. Passes data in only one direction
    C. Passes only ones
    D. Passes only zeros

12. _____ language can be understood by a computer without the need for translation. (Note: All questions in a CET test are multiple choice.)

13. When data goes from the memory to the CPU, the operation is called:

    A. read.
    B. write.

14. Is the output (L) correct for the gate in Fig. C-1?

    A. Yes
    B. No

(Note: Be sure you are able to read timing diagrams before taking any CET test.)

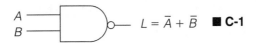
$L = \bar{A} + \bar{B}$ ■ **C-1**

15. In order to transfer data by telephone you need:
    A. serial digital data.
    B. parallel digital data.

16. Which of the following is needed for converting music to pulse-code modulation?
    A. MPC converter
    B. M/C converter
    C. A/D converter
    D. Laser

17. Communication of binary data between computers is accomplished using:
    A. an ampdig.
    B. a modem.
    C. a transponder.
    D. IFF.

18. A quick way to transfer data from one computer into a temporary memory location in another computer is called:
    A. ADM.
    B. AMD.
    C. DMA.
    D. DAM.

19. A single picture element in computer graphics is called a
    _____.

20. Which of the following memories is volatile?
    A. RAM
    B. ROM
    C. PROM
    D. EEPROM

21. A table of logarithms that is permanently stored in a computer for reference is called:
    A. MX table.
    B. ID table.
    C. LAD table.
    D. Look-up table.

327

22. Which of the following accomplishes the same thing as a computer mouse?

    A. Chipmunk
    B. Trap
    C. Trackball
    D. DRAM

23. A memory that stores bits in charged or discharged capacitors is an example of:

    A. A-RAM.
    B. BRAM.
    C. CRAM.
    D. DRAM.

24. Which of the following is a method of checking for errors?

    A. E-Check
    B. Parity Check
    C. 7-Bit Corrector
    D. 8-Bit Corrector

25. What do the letters FORTRAN stand for?

    _____

26. A computer translator that converts a high-level language into machine language is called:

    A. Code Transfer Device (CTD).
    B. Data Reduction Device (DRD).
    C. an inverter.
    D. compiler.

27. Which of the following does not belong in the group?

    A. ADA
    B. COBOL
    C. PASCAL
    D. C-SPACE

28. What is the name for the time between when data is requested and data is received? _____

29. Location of data or, of a program in memory.

    _____

30. Arithmetic and logic operations are carried out in the:

    A. CAD.
    B. CPA.
    C. ALU.
    D. PLA.

31. When groups of data are processed in a group of specified timer intervals it is called:

    A. group processing.
    B. DEM.
    C. batch processing.
    D. program processing.

32. One digit in binary is called a:

    A. bit.
    B. TAB.
    C. BB.
    D. unit.

33. What do the letters CAD/CAM stand for? _____
and _____

34. When the solution of a problem in a computer is expressed, it is called:

    A. carry out.
    B. coding.
    C. decoding.
    D. logging.

35. A group of parallel lines in a computer used for transferring data from one place to another is called the:

    A. control cable.
    B. address cable.
    C. parity loom.
    D. data bus.

36. Roughly a billion bytes is called a:

    A. megabyte.
    B. gigabyte.
    C. multibyte.
    D. mega mega byte.

37. A group of small computers that are interconnected to share data and other resources is called a:

    A. CC.
    B. CDC.
    C. NOC.
    D. LAN.

38. A system of counting based upon 0 through 7 is called a/an _____ system:

    A. hexadecimal.
    B. 4×4.

C. octal.

D. 7-segment.

39. A compiler is a special type of:

    A. hardware.

    B. program.

    C. peripheral device.

    D. I/O device.

40. The letters PLA stand for:

    A. Peripheral LAN Address.

    B. Peripheral LAN Access.

    C. Programmable Logic Array.

    D. Peripheral Logic Array.

41. Refer to the computer block diagram in Fig. C-2. What is the name of the block marked X? _____

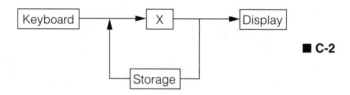

■ C-2

42. The number 12 is represented by 0001 0010 in a numbering system called:

    A. BCD.

    B. Hex.

    C. octal.

    D. reverse polish.

43. Which of the following is used for locating errors?

    A. Half adder

    B. Full adder

    C. Synchroscope

    D. Parity check

44. The circuit in Fig. C-3 is a:

    A. half adder.

    B. full adder.

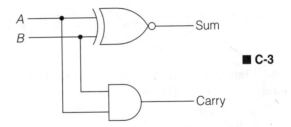

■ C-3

45. What decimal number is represented by binary number 101011?

    A. 37
    B. 43
    C. 47
    D. 534

46. The truth table shown in Fig. C-4 is 2-bit logic gate:

    A. AND.
    B. OR.
    C. NAND.
    D. NOR.

(Note: You should know the truth tables for all of the 2-bit integrated circuit gates.)

331

A	B	L
0	0	0
0	1	0
1	0	0
1	1	1

■ C-4

47. What is the output of the gate circuit in Fig. C-5 when there are two ones at terminals A and B?

    A. $L = 0$
    B. $L = 1$

A ———⊐⊃—— Output   ■ C-5
B ———

48. When a computer is first turned on a _____ program prepares it for operation.

    A. Asynchronous
    B. Bit-by-bit
    C. Lock on
    D. Bootstrap

49. A combination of 8 bits is called a:

    A. nybble (or nibble).
    B. byte.
    C. characteristic.
    D. bode.

50. The section of a computer that interprets and executes instructions is called the:

    A. ACIA.
    B. PIO.
    C. ALU.
    D. converter.

51. Timing pulses that are used to organize computer operations come from the:

    A. ILO.
    B. LO.
    C. PLO.
    D. clock.

52. Two digits can be added, but no carry is generated with a:

    A. NOT carry unit.
    B. partial carrier.
    C. half adder.
    D. full adder.

53. In computer terms, a kilobyte consists of:

    A. 1080 bytes.
    B. 1024 bytes.
    C. 1000 bytes.
    D. 9079 bytes.

54. Is the binary addition problem in Fig. C-6 done correctly?

    A. No
    B. Yes

332

55. A plastic device that must be placed on a tape in order for data to be recorded is called a:

    A. lockout pin.
    B. lockout key.
    C. record pin.
    D. write-enable ring.

56. What is the name of the gate with the following truth table?

A	B	L
0	0	0
0	1	1
1	0	1
1	1	0

57. What is the name of the gate with this truth table?

A	B	L
0	0	1
0	1	0
1	0	0
1	1	1

58. Which of the following is a logic expression for exclusive OR?

    A. $AB + AB = L$
    B. $AB + AB$
    C. Both
    D. Neither

59. Which of the following logic families is known for its very high speed?

    A. TTL
    B. ECL
    C. CMOS
    D. Relay

60. Refer to Fig. C-7. What pin number is marked with an arrow?

61. Which of the following can be used as a data selector?

    A. Multiplexer
    B. Demultiplexer

333

Computer practice test

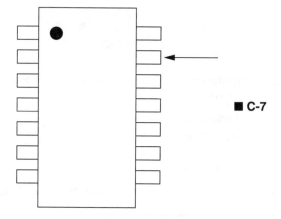

■ C-7

62. Which of the following can be used for serial transmission of data?

    A. Shift register

    B. One-of-seven decoder

    C. D flip-flop

    D. None of these answers is correct.

63. When you multiply 1010_2 by 1011_2, you get:

    A. 1110000_2.

    B. 1101110_2.

    C. 1010101_2.

    D. 1011110_2.

64. Which of the following is a characteristic of random access memories?

    A. Words cannot be retrieved in sequence.

    B. It takes the same amount of time to access any word.

    C. All are nonvolatile.

    D. All of the statements are true.

65. A certain magnetic tape is 2 cm. wide. If the packing density is 100 bits per cm. and there are seven tracks per cm. of tape width, how many bits can be read per second? Assume a tape speed of 20 cm. per second.

    A. 26,000 bits per second

    B. 28,000 bits per second

    C. 2600 bits per second

    D. 260,000 bits per second

334

66. Assume that the disk in Fig. C-8 rotates clockwise. The coded output is:

    A. gray code.
    B. BCD.
    C. binary countdown.
    D. excess 3.

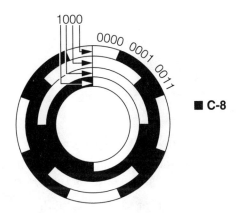

■ C-8

67. Which of the following is not a standard bus?

    A. Control bus
    B. Data bus
    C. Address bus
    D. Arithmetic bus

68. An instruction code register holds the:

    A. previous instruction.
    B. present instruction.
    C. next instruction.
    D. last instruction.

69. The information that is being operated on is stored in a:

    A. data direction register (DDR).
    B. ROM.
    C. accumulator.
    D. memory address register (MAR).

70. Which of the drawings in Fig. C-9 is correct for a memory instruction?

    A. A
    B. B

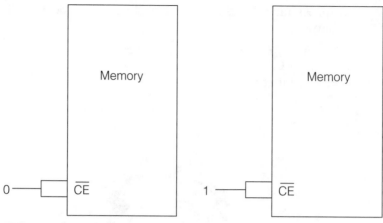

■ C-9

71. Which of the following is correct?

   A. Packet radio is communications between computers.
   B. Packet radio has nothing to do with computers.

72. An optoelectronic decimal light display of the number 8 is normally accomplished with:

   A. 7 segments.
   B. 8 segments.
   C. 9 segments.
   D. 10 segments.

73. Using a CCD, data can be transmitted in a:

   A. serial form.
   B. parallel form.

74. A specific calculation is used to compare two computers. This method of comparing computers is called:

   A. A-B testing.
   B. Benchmark testing.
   C. comptest.
   D. rollover testing.

75. A very fast computer memory in which the most-needed information is stored is called:

   A. fast-track memory.
   B. a memory block.
   C. casche memory.
   D. step-sequence memory.

Answers to practice test

Question	Answer
1.	B
2.	C
3.	B
4.	D
5.	C (Write Once, Read Many)
6.	A
7.	C
8.	C
9.	B
10.	A
11.	A
12.	Machine
13.	A
14.	A
15.	A
16.	C (Analog-Digital Converter)
17.	B
18.	C (Direct Memory Access)
19.	pixel
20.	A
21.	D
22.	C
23.	D
24.	B
25.	FORmula TRANslator
26.	D
27.	D
28.	Access time
29.	Address
30.	C
31.	C
32.	A
33.	Computer-Aided-Design and Computer-Aided-Manufacturing
34.	B
35.	D
36.	B
37.	D
38.	C
39.	B
40.	C

Question	Answer
41.	CPU
42.	A
43.	D
44.	A
45.	B
46.	A
47.	B
48.	D
49.	B
50.	B
51.	D
52.	C
53.	B
54.	B
55.	D
56.	Exclusive OR
57.	Exclusive NOR or coincidence or logic comparator
58.	B
59.	B
60.	13
61.	A
62.	A
63.	B
64.	B
65.	B
66.	C
67.	D
68.	B
69.	C
70.	A
71.	A
72.	A
73.	A
74.	B
75.	C

338

Examples of color codes

D

Diode color codes

THE JEDEC NUMBER HAS A "1N" WITH A SEQUENCE NUMBER of four digits. The four digits are given by four color bands. The first digit is a broad band that also indicates the cathode end. The same color code as used for resistors applies. See Fig. D-1.

The Pro Electron number has three letters and a sequence of two digits. The letters are indicated by two broad bands at the cathode end. The digits are shown by small bands. The color code is given in Table D-1.

■ Table D-1

Broad bands		Small bands
First band	**Second band**	**Serial number**
AA–brown	Z–white	0–black
BA–red	Y–grey	1–brown
	X–black	2–red
	W–blue	3–orange
	V–green	4–yellow
	T–yellow	5–green
	S–orange	6–blue
		7–violet
		8–grey
		9–white

Standard resistance values

E 48 (±2%)	E 96 (±1%)	E 48 (±2%)	E 96 (±1%)	E 48 (±2%)	E 96 (±1%)	E 48 (±2%)	E 96 (±1%)	E 24 (±5%)
1.00	1.00	1.78	17.8	3.16	3.16	5.62	5.62	1.0
	1.02		1.82		3.24		5.76	1.1
1.05	1.05	1.87	1.87	3.32	3.32	5.96	5.90	1.2
	1.07		1.91		3.40		6.04	1.3
1.10	1.10	1.96	1.96	3.48	3.48	6.19	6.19	1.5
	1.13		2.00		3.57		6.34	1.6
1.15	1.15	2.05	2.05	3.65	3.65	6.49	6.49	1.8
	1.18		2.10		3.74		6.65	2.0
1.21	1.21	2.15	2.15	3.83	3.83	6.81	6.81	2.2
	1.24		2.21		3.92		6.98	2.4
1.27	1.27	2.26	2.26	4.02	4.02	7.15	7.15	2.7
	1.30		2.32		4.12		7.32	3.0
1.33	1.33	1.33	2.37	2.37	4.22	7.50	7.50	3.3
	1.37		2.43		4.32		7.68	3.6
1.40	1.40	2.49	2.49	4.42	4.42	7.87	7.87	3.9
	1.43		2.55		4.53		8.06	4.3
1.47	1.47	2.61	2.61	4.64	4.64	8.25	8.25	4.7
	1.50		2.67		4.75		8.45	5.1
1.54	1.54	2.74	2.74	4.87	4.87	8.66	8.66	5.6
	1.58		2.80		4.99		8.87	6.2
1.62	1.62	2.87	2.87	5.11	5.11	9.09	9.09	6.8
	1.65		2.94		5.23		9.31	7.5
1.69	1.69	3.01	3.01	5.36	5.36	9.53	9.53	8.2
	1.74		3.00		5.49		9.76	9.1

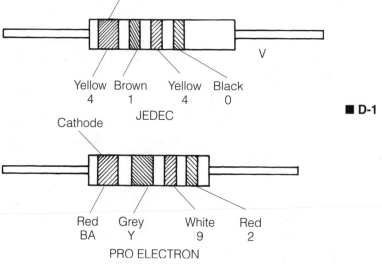

Cathode

Yellow Brown   Yellow Black
4   1   4   0

V

JEDEC

■ D-1

Cathode

Red   Grey   White   Red
BA   Y   9   2

PRO ELECTRON

Axial lead

Black	– 0	Black	– 0
Brown	– 1	Brown	– 1
Red	– 2	Red	– 2
Orange	– 3	Orange	– 3
Yellow	– 4	Yellow	– 4
Green	– 5	Green	– 5
Blue	– 6	Blue	– 6
Violet	– 7	Violet	– 7
Gray	– 8	Gray	– 8
White	– 9	White	– 9

Reliability code

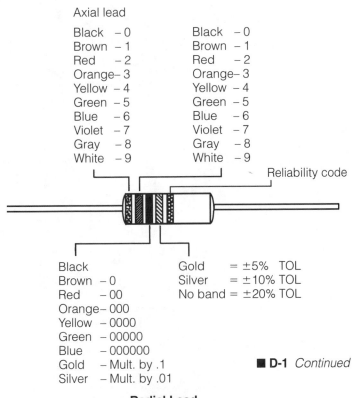

Black		Gold = ±5% TOL
Brown	– 0	Silver = ±10% TOL
Red	– 00	No band = ±20% TOL
Orange	– 000	
Yellow	– 0000	
Green	– 00000	
Blue	– 000000	
Gold	– Mult. by .1	
Silver	– Mult. by .01	

■ **D-1** *Continued*

341

Radial Lead

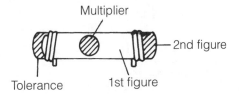

Multiplier

Tolerance   1st figure   2nd figure

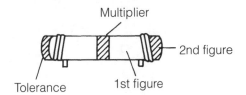

Multiplier

Tolerance   1st figure   2nd figure

Example:
Color code: Red, Violet, Orange, Gold
27,000 Ω ± 5%
Color code: Brown, Black, Black, Gold
10 Ω ± 5%

Examples of color codes

Laws of Boolean algebra

$$1 \times 1 = 1 \qquad\qquad 1 + 1 = 1$$
$$1 \times 0 = 0 \qquad\qquad 1 + 0 = 1$$
$$A \times 1 = A \qquad\qquad A + 1 = 1$$
$$A \times 0 = 0 \qquad\qquad A + 0 = A$$
$$A \times A = A \qquad\qquad A + A = A$$
$$\overline{A} \times A = 0 \qquad\qquad \overline{A} + A = 1$$

An even number of overbars is the same as no overbars:

$$\overline{\overline{A}} = A \qquad \overline{\overline{\overline{\overline{A}}}} = A$$

An odd number of overbars is the same as one overbar:

$$\overline{\overline{\overline{A}}} = \overline{A} \qquad \overline{\overline{\overline{\overline{\overline{A}}}}} = \overline{A}$$

The basic rules of algebra apply.

Example: $A \times B + A \times C = A(B + C)$.

DeMorgan's Theorem: Break the bar and change the sign. For example:

$$\overline{A \times B} = \overline{A} + \overline{B}$$
$$\overline{A + B} = \overline{A} \times \overline{B}$$

List of acronyms

A	ampere, amplification, address
Av	voltage amplification
ACC	automatic chrominance control
A/D	analog/digital
ACIA	asynchronous communications interface adaptor
AFC	automatic frequency control
AFT	automatic fine tuning
AGC	automatic gain control
ALU	arithmetic logic unit
AM	amplitude modulation
ASCII	American Standard Code for Information Exchange
b, B	base, bandwidth
B	beta
BCD	binary-coded decimal
C	capacitance, collector
CAD	computer assisted aided design
CAM	computer aided machining
CATV	community antenna television, cable television
CCD	charge coupled device
CET	certified electronic technician
CMOS	complementary metal-oxide semiconductor
CMRR	common-mode rejection ratio
COBOL	COmmon Business Oriented Language
CPU	central-processing unit
CTD	code-transfer device
CTL	complementary-transistor logic
D	Drain, data, density, distance
D/A	digital analog
DTD	data-transfer device
dB	decibel
DIAC	DIode AC switch

DL	delay line
DMA	direct memory access
e, E	emitter
ECL	emitter-coupled logic
EEPROM	electrical erasable programmable read-only memory
ELP	extended long play
f, F	freqeuency
FET	field-effect transistor
FG	frequency-generator
FM	frequency-modulation
FORTRAN	FORmula TRANslation
h, H,	height, horizontal, hertz
HOT	Horizontal output transformer
I	current
IC	integrated circuit
IF	intermediate frequency
IRQ	interrupt request
IRE	Institute of Radio Engineers
ISCET	International Society of Certified Electronics Technicians
j	a math operator (90 degrees)
k	a constant, kilo (1000)
L	length, inductance
LAD	light-activated diode
LAN	local area network
LED	light-emitting diode
LDR	light-dependant resistor
LP	long-play
m	milli
M	meg
MOSFET	metal-oxide semiconductor field-effect transistor
MX	multiplex
n, N	negative, north, number
NOT IRQ	not interrupt request
N-S	North-South
NMI	nonmaskable interrupt

NTSC	National Television Systems Committee
op amp	operational amplifier
p, P	positive, power
PAL	phase alternation line
PG	pulse generator
pi	3.1416
PLL	phase-locked loop
PROM	programmable read-only memory
Q	quality, figure of merit
R	resistance, resistor
RAM	random-access memory
RF	radio frequency
R/NOT W	read not/write
ROM	read-only memory
S	source
SCA	subsidiary communications authorization
SCR	silicon-controlled rectifier
SLP	super long play
S-N	south-north
SP	standard play
t, T	time, temperature , period
TAB	Technical Author's Bureau
U	ultra
UHF	ultra-high frequency
UJT	unijunction transitor
v, V	voltage, vertical
VCO	voltage-controlled oscillator
VCR	video-cassette recorder
VDR	voltage-dependant resistor
VHF	ultra-high frequency
VIRS	vertical -interval reference signal
VITS	vertical-interval test signal
VMA	valid memory address
VRT	video tape recorder
Xc	capacitive reactance

Xl inductive reactance

Z impedance

Index

A

A/D converters, 148
accumulators, 143
active circuits, 49
active filters, 94, **94**
alpha cutoff frequency, 63, 81
amplifiers, 54, 56, **55**
  bandwidth of, 63
  beta-squared, 58-59
  burst, 156
  Class A, 62
  complementary (*see* amplifiers, totem-pole)
  configurations, 81
  coupled, 75-78, **76**
  Darlington, 58-59, 60, 242, **59**
  differential, 84-85, **59**, **89**
  distortion in, 62-63
  equations for, 81
  inverting, 88
  Loftin-White, 76
  noise in, 60-61
  noninverting, 91-92, **91**
  operational, 79-82, 84, 88-92, 234
  stacked, **59**
  sync, 156
  testing, 200-203
  totem-pole, 60, 87, **59**
  voltage gain, 64
AND, 117
angular velocity, 22
arc sin, 195
arithmetic logic unit (ALU), 138
Armstrong oscillator, 101
arrays, resistor, 17
asynchronous circuits, 146
asynchronous counters, 147
attenuators, 52-53
automatic fine tuning (AFT), 157
automatic frequency control (AFC), 155, 156

automatic gain control (AGC), 155, 157-158
autotransformers, 25
average picture level (APL), 174

B

baluns, 25
bandwidth, 63, 83, 239
batteries (*see also* power supplies)
  parallel connection, 97, **97**
  primary cells, 97
  secondary cells, 97
beat frequency oscillator, 244, 245
beta cutoff frequency, 63, 81
beta-squared amplifiers, 58-59
bifilar winding, 17
binary number system, 35, 113
  converting to/from decimal, 115
binary-coded decimal number system, 113
bipolar junction transistor (BJT), 81
bipolar transistors, 58, 199, **58**
bit slice, 144-145
bits, 138-139
bode plot, 63, 82, **82**
Boolean algebra, 116-117, 343
bootstrap circuit, 57, **57**
bridges
  Maxwell, 234
  Shearing, 234
  Wein, 234
  Wheatstone, 234, 242
bruch sequencer, 174
buffers, 91
  tri-state, 117, 127, **126**
burst separator, 156
burst signal, 156

C

capacitance, 18, 236
capacitive reactance, 21

Illustrations are indicated in **boldface.**

capacitors, 18-21, 27, 49-50
  ceramic, 19
  electrolytic, 19-20, 246
  mica, 19
  mylar-film, 19
  paper, 18
  variable, 20
carbon resistors, 14-16
cascaded, 75
cascoded, 75
cathodes, **64**
cells
  primary, 97
  secondary, 97
ceramic capacitors, 19
certified electronics technician (*see* CET)
CET, xi-xii
  Associate-level, 1
    certificate, **2**
  Journeyman-level, 1
    certificate, **1**
CET test, 1-11
  answer sheets, **6**, **7**
  Associate-level
    practice, 275-305
    section headings, 2-4
  certificates, 1, **1**, **2**
  educated guessing, 214
  Journeyman-level, section headings, 4-5, 8
  pacing yourself, 213-214
  practice, 215-246
    associate-level, 275-305
    computer, 325-338
    consumer products, 307-323
    timed, 248-274
  ratings, 1
  rules for taking, 8-9
  time limit, 247-248
charge coupling, 148
charge-coupled device (CCD), 148
chrominance, 174
chrominance signal, 174-175
circuit breakers, 101
circuits, 13
  active, 49
  analog, 75-112
  asynchronous, 146
  bilateral, 13
  bootstrap, 57, **57**
  capacitors in, 18-21
  color killer, 159
  crowbar, 101, **102**

electronic, 49-53
  in parallel, 50-51, **51**
  in series, 50, **51**
  inductors in, 21-24
  integrated, 79-82
  linear, 13
  passive, 49
  protection, 101
  quizzes/answers, 38-47, 52-53, 64-73, 102-112
  resistors in, 14
  safety, 53, **53**
  sample-and-hold, 93
  sense, 98
  synchronous, 146
Class A amplifier, 62
clipping, 62
color bar, 175
color burst, 151, 175
color codes, 339-341
color killer circuit, 159
color subcarrier, 175
Colpitts oscillator, 101
comb filter, 158-159
common base, 56-57
common collector, 56-57, 271
common emitter, 56-57
common gain, 56
common gate, 56
common-mode rejection ratio (CMRR), 86
common source, 56
companders, 93
comparators, 92
  logic, 117
complementary amplifiers (*see* totem-pole amplifiers)
complex number system, 35
complimentary metal oxide semiconductor (CMOS), 136, 146
composite video signal, 175
constant-current diodes, 28-30, 239, **29**, **31**
continuous wave, 245
control track logic (CTL), 173
converters, 101
  A/D, 148
  D/A, 148
correction voltage, 94
counter electromotive force (CEMF), 24
counter voltage, 24
counters
  asynchronous, 147
  programmable, 147

350

ripple, 147
synchronous, 147
coupling
  amplifiers, 75-78, **76**
  charge, 148
  direct, 76
  impedance, 78
  resistor-capacitor, 76
  transformer, 78
cross modulation, 62
crossover distortion, 203
crowbar circuit, 101, **102**
current regulator, 99, **99**

D

D flip-flops, 134-135, **135**
D/A converters, 148
Darlington amplifiers, 58-59, 60, 242, **59**
data latch, 131
decibels, 244
decimal number system, 35, 113
  converting to/from binary, 115
decoders, 147-148
decoupling filter, 78
degeneration, 83
DeMorgan's theorems, 124
demultiplexers, 94, 139
diacs, 26, **27**
diamagnetic materials, 232
differential amplifiers, 84-85, **59**, **89**
differential gain, 175
differential phase, 175
digital systems, 113-148
diodes, 26
  constant-current, 28-30, 239, **29**, **31**
  four-layer, 32
  hot-carrier, 32, **33**
  in parallel, 27, **28**
  light-activated, 31
  light-emitting, 31
  optoelectronic, 31-32, **32**
  rectifier, 27-28, **28**
  Schottky, 32
  Shockley, 32
  tunnel, 31, **31**
  vacuum-tube, **80**
  varactor, 33, **33**
  zener, 33, **31**, **33**
DIP package, 124-125
distortion
  crossover, 203
  harmonic, 62

in amplifiers, 62-63
intermodulation, 62
phase-shift, 63
dynamic logic systems, 129
dynamic memory, 148

E

electrical units, 35
electrically erasable programmable read-only
    memory (EEPROM), 139
electrolytic capacitors, 19-20, 246
electromotive force (EMF), 24
emitter base, 272
emitter-coupled logic (ECL), 136, 147
encoders, 147-148
equivalent series resistance (ESR), 20
erasable programmable read-only memory
    (EPROM), 139
exclusive NOR, 117
exclusive OR, 117
expanders, 93

F

Faraday shield, 236
feedback resistance, 90
ferromagnetic materials, 231
field effect transistor (FET), 199
film resistors, 16
filters, 49
  active, 94, **94**
  comb, 158-159
  decoupling, 78
  four-terminal, 49-50, **50**
  low-pass, **94**
  passive, 94
flip-flops, 129-137
  D, 134-135, **135**
  J-K, 135-137
  RS, 131-133
  toggled, 133-134, **134**
flyback transformer, 157
FM signals, 153, **154**
  televisions and, 149-156
frequency, 63
  intermediate, 149
  radio, 149
  video, 161, **161**
fuses, 101

G

gain
  differential, 175

gain *continued*
  power, 64
  voltage, 64
gain-bandwidth, 63, 83
galvanometer, 188
gen-lock, 175
glitches, 146
gyrators, 25, 92

H

H rate, 175
half adder, 117
Hall sensor, 230-231
harmonic distortion, 62
Hartley oscillator, 101
helical scanning, 165, **165**, **166**
heterodyning, 149-150
hexadecimal number system, 113
horizontal output transformer (HOT), 156
hot-carrier diodes, 32, **33**
hue, 175

I

impedance coupling, 78
impedance matching, 37-38, **38**
inductive reactance, 22, 35
inductors, 21-24, 50
input resistance, 90
insertion loss, 53
Institute of Electrical and Electronics
  Engineers (IEEE), 35
integrated circuits (ICs), 79-82
intermediate frequency (IF), 149
intermodulation distortion, 62
International Society of Certified Electronics
  Technicians (ISCET), xii
Inv sin, 195
inverters, 101, 117
  tri-state, 117, 127, **126**
inverting amplifiers, 88

J

j operators, 35
J-K flip flops, 135-137
Johnson noise, 14, 61
Joule, 241
junction field effect transistor (JFET), 82

K

kilowatt-hour, 241

L

lamps, 130
laws, Ohm's, 13
level shifting, 76
light-activated diode (LAD), 31
light-emitting diode (LED), 31
Lissajous pattern, 194-196
load, 99
load resistance, 99
Loftin-White amplifiers, 76
logic comparators, 117
logic gates, 117-123, **118**, **119**, **120**, **121**
  AND, 117
  exclusive NOR, 117
  exclusive OR, 117
  inverter, 117
  NAND, 117, 131, **131**
  NOR, 117
  NOT, 117
  OR, 117
  tri-state buffers, 117
  tri-state inverters, 117
  truth tables 121-122, **120**, **131**
logic systems
  dynamic, 129
  static, 129
low-frequency compensating network, 77-78,
  77
low-pass filters, **94**
luminance, 175

M

magnetrons, 26
master oscillator, 244
Maxwell bridge, 234
memory, 137-140
  dynamic, 148
  EEPROM, 139
  EPROM, 139
  PROM, 139
  RAM, 139
  ROM, 139
  static, 148
  volatile, 273
metal-oxide semiconductor field-effect
  transistor (MOSFET), 61, 82, **80**
metal-film resistors, 16
mica capacitors, 19
microprocessors, 137-140, **140**
  bits, 138-139
  dedicated, 138

pins, 140-143
  undedicated, 138
modulation, cross, 62
motors, stepping, 232
multiplexers, 94, 139
multipliers, 188
mylar-film capacitors, 19

N

NAND, 117, 131, **131**
nanofarads, 35
National Television Standards Committee
    (NTSC), 175
neon lamps, 26, **27**
nepers, 244
networks, low-frequency compensating, 77-
    78, **77**
noise
  in amplifiers, 60-61
  Johnson, 14, 61
  partition, 60-61
  thermal agitation, 14, 61
noninverting amplifiers, 91-92, **91**
nonlinear resistors, 33-34
nonsinusoidal oscillator, 101
NOR, 117
NOT, 117
number systems
  binary, 35, 113
  binary-coded decimal, 113
  binary-to-decimal conversion, 115
  complex, 35
  decimal, 35, 113
  decimal-to-binary conversion, 115
  hexadecimal, 113
  octal, 113
  radix values, 113-114

O

octal number system, 113
Ohm's law, 13
ohms, 246
op amps, 88-92, 234
  characteristics, 84
  IC, 79-82
operational amplifiers (*see* op amps)
OR, 117
oscillators, 101
  Armstrong, 101
  beat frequency, 244, 245
  Colpitts, 101
  Hartley, 101

master, 244
nonsinusoidal, 101
sine-wave, 101
sinusoidal, 101
voltage-controlled, 94
oscilloscopes, 188, 193-195
  triggered-sweep, 196
overbars, 123

P

padders, 21
pads, 52-53
paper capacitors, 18
paramagnetic materials, 232
parasitics, 243
partition noise, 60-61
passive circuits, 49
passive filters, 94
pentrodes, **80**
percent regulation, 98
phase alternation line (PAL), 175
phase locked loop (PLL), 94-96, 144, 157, **95**
phase-shift distortion, 63
phons, 244
picofarads, 35
polarity, 54
potentiometers, 17
power, RMS, 245
power gain, 64
power rating, 14
power supplies, 97-101
  analog regulated, 98-101, **99**
  scan-derived, 158
  television, 158, **159**
power transformer, 158
primary cells, 97
probes, 197
programmable counters, 147
programmable read-only memory (PROM),
    139
programmable unijunction transistor (PUT),
    30

Q

quizzes and answers
  CET practice test, 215-246
    associate-level, 275-305
    computers, 325-338
    consumer products, 307-323
    timed, 248-274
  CET test, 10-11
  circuits, 38-47, 64-73, 102-112

quizzes and answers *continued*
  mathematics, 38-47
  television and VCRs, 176-186
  test equipment and testing, 203-211

R

radio frequency (RF), 149
random-access memory (RAM), 139
raster, 157
reactance
  capacitive, 21
  inductive, 22, 35
read-only memory (ROM), 139
  electrically erasable programmable, 139
  erasable programmable, 139
  programmable, 139
rectifiers, 27-28, 101, **28**
  silicon-controlled, 101
registers, 143-144, **144**
regulation, 98
regulators
  current, 99, **99**
  switching, 100, **100**
  tracking, 102
relays, 130
resistance, 14, 21, 35
  equivalent series, 20
  feedback, 90
  input, 90
  load, 99
resistor arrays, 17
resistor-capacitor coupling, 76
resistors, 14, 27
  carbon, 14-16
  film, 16
  nonlinear, 33-34
  surge-limiting, 271
  variable, 17-18, **18**
  voltage-dependent, 24, 33-34, 241, **34**
  wire-wound, 16-17, **17**
resonant frequency, 50
response curve, 82
rheostats, 17
ringing test, 197-198
ripple counters, 147
RMS power, 245
RS flip-flops, 131-133

S

safety circuit, 53, **53**
sample-and-hold circuits, 93
saturation, 176

sawtooth waveform, 30, **30**
scan-derived power supply, 158
Schottky diodes, 32
secondary cells, 97
sense circuit, 98
sense voltage, 98
servo systems, 172-174, **173**
Shearing bridge, 234
Shockley diodes, 32
shunt, 188
siemens, 246
silicon-controlled rectifier (SCR), 101
sine-wave oscillators, 101
sinusoidal oscillator, 101
slew rate, 87-88
somes, 244
square waves, 197
staircase, 176
static logic systems, 129
static memory, 148
stepping motors, 232
summing point, 90
surge-limiting resistor, 271
switches, 130, 232
switching regulator, 100, **100**
sync amplifier, 156
sync separator, 156
synchronous circuits, 146
synchronous counters, 147

T

television
  color, 153, **155**
  color burst, 151
  FM signals and, 149-156
  frequencies, **150**
  monochrome, 152-153, **154**
  pincushion module, 159-160
  power supply, 158, **159**
  quizzes/answers, 176-186
  receiver, 154-156
  signal amplitudes, **151**
  terms regarding, 174-176
  vertical interval reference signal (VIRS), 152
  vertical interval test signal (VITS), 152
test equipment
  galvanometer, 188
  meter movements, 188-193
  oscilloscopes, 188, 193-195
  probes, 197
  quizzes/answers, 203-211

354

volt-ohm-milliammeter, 198
voltmeter, 189-190
tests and testing (*see also* CET tests; quizzes and answers)
    amplifiers, 200-203
    evaluating parameters, 198
    ringing, 197-198
    square-wave, 197
tetrodes, **80**
thermal agitation noise, 14, 61
thermistors, 33-34, **34**
thyratrons, **80**
thyristors, 28, 64, **64**
time constants, 22-24, **23**
timers, 93, **93**
toggled flip-flops, 133-134, **134**
totem-pole amplifiers, 60, 87, **59**
tracking regulators, 102
transconductance, 82
transformer coupling, 78
transformers, 25-26
    auto-, 25
    dot notation for, 25, **25**
    flyback, 157
    horizontal output, 156
    power, 158
transient voltage, 236
transistors, 130, 202-203
    bipolar, 58, 199, **58**
    bipolar junction, 81
    field effect, 199
    junction field effect (JFET), 82
    metal oxide semiconductor field effect (MOSFET), 61, 82, **80**
    NPN, 54, 56
    programmable unijunction (PUT), 30
    unijunction, 28-30, **29**
transistor-transistor logic (TTL), 135-136, 146
tri-state devices, 126-129
    buffers, 127, **126**
    inverters, 127, **126**
trimmers, 21
triodes, **80**
truth tables, 121-122, **120**, **131**
tunnel diodes, 31, **31**
twiggered-sweep oscilloscopes, 196

U

unijunction transistor (UJT), 28-30, **29**
units, electrical, 35

V

vacuum-tube diode, **80**
vacuum tubes, 199
varactor diodes, 33, **33**
variable capacitors, 20
variable resistors, 17-18, **18**
vertical interval reference signal (VIRS), 152
vertical interval test signal (VITS), 152, 176
video frequencies, 161, **161**
video head, 161, **162**
video home system (VHS), 160, 165-167
video spectrum, 164, **164**
videocassette recorders (VCRs), 160-164
    block diagram, 167, 170-172, **168**, **169**
    capstan, 173
    control track logic, 173
    drum, 174
    frequency generator sensor, 173
    heads, 161, **162**
    helical scanning, 165, **165**, **166**
    quizzes/answers, 176-186
    servo systems, 172-174, **173**
    VHS, 160, 165-167
    wavelength vs. head gap, 163, **163**
virtual ground, 90
volatile memory, 273
volt-ohm-milliammeter (VOM), 198
voltage, 35, 37, **36**
    correction, 94
    counter, 24
    sense, 98
    transient, 236
voltage-controlled oscillator (VCO), 94
voltage-dependent resistor (VDR), 24, 33-34, 241, **34**
voltage gain, 64
voltage regulators, **80**
voltmeter, 189-190

W

watt-second, 241
waveforms, 238
    sawtooth, 30, **30**
    square, 197
Wein bridge, 234
Wheatstone bridge, 234, 242
winding, bifilar, 17
wire-wound resistors, 16-17, **17**

Z

zener diodes, 33, **31**, **33**

355

About the authors

Joe Risse

Joseph A. "Joe" Risse has worked as an assembler, technician, engineer, maintenance engineer, transmitter operator, chief broadcast engineer, director of electronics department for a correspondence school, project manager for industrial training programs, and is active on advisory committees for a technical institute and vo tech school. He has completed courses in the military, college courses in electronics, correspondence school courses, and industrial group/training programs.

He holds the B.A. degree in Natural Science/Mathematics from Thomas Edison College, is a Fellow of the Society of Broadcast engineers and the Radio Club of America, a Life Member of the Electronic Technicians Association, and Member of the International Society of Certified Electronics Technicians. He is a registered Professional Engineer by the Commonwealth of Pennsylvania, and certified as an electronics technician by both ISCET and ETA.

Risse has completed all of the electronics, mathematics, and physics courses required for Electronics major in Physics degree at the University of Scranton. He has prepared certification and practice exams for ISCET. He was the editor of the *Journal of the Society of Broadcast Engineers* during the early years of the SBE, and held the national office of executive vice president for 2 years.

Sam Wilson

Sam Wilson earned his bachelor's degree from Long Beach State College and his master's degree from Kent State University. He also has diplomas from Capitol Radio Engineering Institute and RCA Institutes.

Wilson is now a full-time technical writer and consultant. In 1983, he was selected as *Technician of the Year* by the International Society of Certified Electronics Technicians (ISCET). He has been the CET Test Consultant for that organization; and, has been the Technical publications Director for the National Electronics Service Dealers Association (NESDA).

His electronics experience includes 18 years as an instructor and professor. He also has 12 years of practical experience as technician and engineer. He is presently a specialist in training equipment design and preparation of technical training publications. Over the years, Wilson has written 27 technical electronics books.